# INTERNATIONAL
# WILDLIFE
## ENCYCLOPEDIA

## THIRD EDITION

## Volume 18
## SOL–SWA

MARSHALL CAVENDISH
New York • London • Toronto • Sydney

Marshall Cavendish Corporation
99 White Plains Road
Tarrytown, New York 10591–9001

Website: www.marshallcavendish.com

Library of Congress Cataloging-in-Publication Data

Burton, Maurice, 1898-
    International wildlife encyclopedia / [Maurice Burton, Robert Burton] .-- 3rd ed.
        p. cm.
    Includes bibliographical references (p.).
    Contents: v. 1. Aardvark - barnacle goose -- v. 2. Barn owl - brow-antlered deer -- v. 3. Brown bear - cheetah -- v. 4. Chickaree - crabs -- v. 5. Crab spider - ducks and geese -- v. 6. Dugong - flounder -- v. 7. Flowerpecker - golden mole -- v. 8. Golden oriole - hartebeest -- v. 9. Harvesting ant - jackal -- v. 10. Jackdaw - lemur -- v. 11. Leopard - marten -- v. 12. Martial eagle - needlefish -- v. 13. Newt - paradise fish -- v. 14. Paradoxical frog - poorwill -- v. 15. Porbeagle - rice rat -- v. 16. Rifleman - sea slug -- v. 17. Sea snake - sole -- v. 18. Solenodon - swan -- v. 19. Sweetfish - tree snake -- v. 20. Tree squirrel - water spider -- v. 21. Water vole - zorille -- v. 22. Index volume.
    ISBN 0-7614-7266-5 (set) -- ISBN 0-7614-7267-3 (v. 1) -- ISBN 0-7614-7268-1 (v. 2) -- ISBN 0-7614-7269-X (v. 3) -- ISBN 0-7614-7270-3 (v. 4) -- ISBN 0-7614-7271-1 (v. 5) -- ISBN 0-7614-7272-X (v. 6) -- ISBN 0-7614-7273-8 (v. 7) -- ISBN 0-7614-7274-6 (v. 8) -- ISBN 0-7614-7275-4 (v. 9) -- ISBN 0-7614-7276-2 (v. 10) -- ISBN 0-7614-7277-0 (v. 11) -- ISBN 0-7614-7278-9 (v. 12) -- ISBN 0-7614-7279-7 (v. 13) -- ISBN 0-7614-7280-0 (v. 14) -- ISBN 0-7614-7281-9 (v. 15) -- ISBN 0-7614-7282-7 (v. 16) -- ISBN 0-7614-7283-5 (v. 17) -- ISBN 0-7614-7284-3 (v. 18) -- ISBN 0-7614-7285-1 (v. 19) -- ISBN 0-7614-7286-X (v. 20) -- ISBN 0-7614-7287-8 (v. 21) -- ISBN 0-7614-7288-6 (v. 22)
    1. Zoology -- Dictionaries.    I. Burton, Robert, 1941-    . II. Title.

    QL9 .B796 2002
    590'.3--dc21

                                                            2001017458

Printed in Malaysia
Bound in the United States of America

07 06 05 04 03        8 7 6 5 4 3

Brown Partworks
Project editor: Ben Hoare
Associate editors: Lesley Campbell-Wright, Rob Dimery, Robert Houston, Jane Lanigan, Sally McFall, Chris Marshall, Paul Thompson, Matthew D. S. Turner
Managing editor: Tim Cooke
Designer: Paul Griffin
Picture researchers: Brenda Clynch, Becky Cox
Illustrators: Ian Lycett, Catherine Ward
Indexer: Kay Ollerenshaw

Marshall Cavendish Corporation
Editorial director: Paul Bernabeo

## Authors and Consultants

Dr. Roger Avery, BSc, PhD (University of Bristol)

Rob Cave, BA (University of Plymouth)

Fergus Collins, BA (University of Liverpool)

Dr. Julia J. Day, BSc (University of Bristol), PhD (University of London)

Tom Day, BA, MA (University of Cambridge), MSc (University of Southampton)

Bridget Giles, BA (University of London)

Leon Gray, BSc (University of London)

Tim Harris, BSc (University of Reading)

Richard Hoey, BSc, MPhil (University of Manchester), MSc (University of London)

Dr. Terry J. Holt, BSc, PhD (University of Liverpool)

Dr. Robert D. Houston, BA, MA (University of Oxford), PhD (University of Bristol)

Steve Hurley, BSc (University of London), MRes (University of York)

Tom Jackson, BSc (University of Bristol)

E. Vicky Jenkins, BSc (University of Edinburgh), MSc (University of Aberdeen)

Dr. Jamie McDonald, BSc (University of York), PhD (University of Birmingham)

Dr. Robbie A. McDonald, BSc (University of St. Andrews), PhD (University of Bristol)

Dr. James W. R. Martin, BSc (University of Leeds), PhD (University of Bristol)

Dr. Tabetha Newman, BSc, PhD (University of Bristol)

Dr. J. Pimenta, BSc (University of London), PhD (University of Bristol)

Dr. Kieren Pitts, BSc, MSc (University of Exeter), PhD (University of Bristol)

Dr. Stephen J. Rossiter, BSc (University of Sussex), PhD (University of Bristol)

Dr. Sugoto Roy, PhD (University of Bristol)

Dr. Adrian Seymour, BSc, PhD (University of Bristol)

Dr. Salma H. A. Shalla, BSc, MSc, PhD (Suez Canal University, Egypt)

Dr. S. Stefanni, PhD (University of Bristol)

Steve Swaby, BA (University of Exeter)

Matthew D. S. Turner, BA (University of Loughborough), FZSL (Fellow of the Zoological Society of London)

Alastair Ward, BSc (University of Glasgow), MRes (University of York)

Dr. Michael J. Weedon, BSc, MSc, PhD (University of Bristol)

Alwyne Wheeler, former Head of the Fish Section, Natural History Museum, London

## Picture Credits

# Contents

# SOLENODON

THE FOUR SPECIES OF solenodons were first discovered by Europeans in 1833 on the islands of Hispaniola and Cuba. Fossil evidence shows that these shrewlike creatures were in existence 30 million years ago in North America. They seem therefore to be primitive insectivores, survivors from the past that have managed to survive in these two islands because there are so few natural enemies there.

Solenodons are 11–15 inches (28–39 cm) long, with a tail 6½–10 inches (17–26 cm) long. They have a stout body with a large head and a long snout, which usually forms a cushion of flesh just in front of the tips of the nasal bones. Solenodons have a rod of cartilage in front of the nasal bones, supporting the long snout. There are many long bristles on the face, the eyes are very small and the ears are partly naked and mostly hidden in the fur. The coat is blackish to reddish brown, paler on the underside. The tail is nearly naked and so are the legs and the large feet, each with five toes bearing large, strong claws. The especially large forefeet have larger and more curved claws than those on the hind feet.

Solenodons are nocturnal. During the day they lie up in burrows, in hollow trees and logs or in caves, well out of sight. When they do come out, they run on their toes with a stiff waddle, following an erratic, almost zigzag course. Like some shrews, solenodons have poisonous saliva, and the second incisor on each side in the lower jaw is grooved (indeed, the word solenodon means "grooved tooth"). At the base of each of these incisors is a gland from which the poison runs along the groove in each tooth. When captive solenodons fight, the light wounds inflicted are often fatal.

Solenodons feed on insects, worms, small invertebrates, small reptiles, roots, fruits and leaves. They root in the ground with their long snouts, dig with their stout claws or rip open rotten logs. Captive solenodons bathe often and drink only when bathing. Although food sources are plentiful, solenodons are becoming rarer. This is thought to be due mainly to deforestation, increased human activity and predation from introduced dogs and cats.

*Solenodons are not closely related to shrews, although there is a definite physical resemblance. Pictured is a Haitian solenodon, Solenodon paradoxus.*

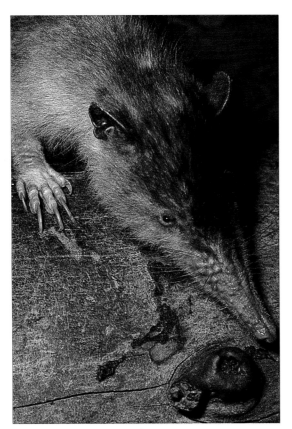

## SOLENODONS

| | |
|---|---|
| CLASS | **Mammalia** |
| ORDER | **Insectivora** |
| FAMILY | **Solenodontidae** |
| GENUS AND SPECIES | ***Solenodon cubanus;*** ***S. paradoxus; S. arredondoi; S. marcanoi*** |

WEIGHT
**Up to 2⅕ lb. (1 kg)**

LENGTH
**Head and body: 11–15 in. (28–39 cm);
tail: 6½–10 in. (17–26 cm)**

DISTINCTIVE FEATURES
**Resemble large shrews; very long bare nose; tiny eyes; long, well-built legs; large feet and claws; blackish to reddish brown coat**

DIET
**Invertebrates; fruits and other vegetable matter; small reptiles**

BREEDING
**Breeding season: irregular and aseasonal; number of young: usually 1, rarely 2; gestation period: not known**

LIFE SPAN
**Up to 11 years in captivity**

HABITAT
**Brushland, forests, often plantation margins**

DISTRIBUTION
**Hispaniola and Cuba**

STATUS
***S. cubanus*** **and** ***S. paradoxus:*** **endangered;** ***S. arredondoi*** **and** ***S. marcanoi:*** **sometimes classed as extinct**

☐ Solenodons

# SOLIFUGID

THESE RELATIVES OF SPIDERS have among the most powerful jaws, in proportion to body size, of any animal. They are known by a wide range of common names, including false spiders, wind scorpions, sun spiders and camel spiders, so most zoologists refer to them collectively as solifugids, or occasionally as solpugids. The order to which they belong is known as the Solifugae or the Solpugida, and it is placed in the class Arachnida, along with spiders, ticks, scorpions and other eight-legged invertebrates.

A solifugid looks a little like a hairy spider. Like a spider, its body is made up of two major parts. The forepart is the prosoma, or cephalothorax, with eight segments; the hind part, with 10 segments, is the abdomen. The prosoma is enlarged at the front to accommodate the powerful chelicerae (jaws) and the large muscles needed to work them. The pincerlike chelicerae are held up high while the animal is moving and are used to seize prey.

## Purposeful build

The solifugid appears to have five pairs of legs, but the first pair are in fact pedipalps. These are long, powerful and very hairy, and each has a sucker at the end, useful in climbing smooth surfaces. The pedipalps also serve as sensory organs and in seizing and manipulating prey. Behind them are four pairs of walking legs, each tipped with two sharp claws, although the first pair are normally held up off the ground, like antennae. On each of the two hindmost legs are five racquet-shaped organs, known as malleoli. These are thought to detect the odors of prey or potential mates. Sensory hairs all over the body assist in detecting movement.

The 800 solifugid species are usually yellow, brown or black. A few have striped bodies. Body length is in the range of ½–3 inches (1.2–7.5 cm), and the larger species have a 6-inch (15-cm) legspan. They are especially numerous in desert regions of Africa and Arabia, and occur as far east as Vietnam. In Europe they occur only in southeastern Spain. In North America they are found in Florida, in the deserts of the southwestern United States and around hot springs as far north as southwestern Canada. They also live in Central America, parts of Brazil and coastal areas of northern and western South America.

## Sand plow

The name solifugid means "fleeing from the sun," and most of the species are nocturnal. Some, however, are active by day, including the *aranhos del sol* (Spanish for "sun spiders"). These all run quickly, and the diurnal solifugids have been likened to balls of thistledown blown over the sand as they run about, hence the name wind scorpions. From time to time they suddenly stop running and quarter the ground, probably to track the scent of prey. With their active lifestyle, solifugids need an efficient breathing system; instead of the book lungs of spiders, they have a tracheal system, like that of insects.

The nocturnal species hide under stones by day or bury themselves in the sand using their stout jaws as plows or scraping the sand backward with their second pair of legs. If attacked, solifugids rock back and forth and stridulate menacingly with their jaws. When confronted by scorpions, solifugids tend to push the opponent away with the jaws and then run away, though if they stand and fight they may try to snip off the scorpion's stinger.

## Pulped prey

A solifugid kills and eats insects, scorpions, spiders, and smaller solifugids, as well as small mice, birds and lizards. Once its prey is subdued,

*A cricket is easy prey for this giant solifugid on arid savanna in South Africa. Long legs, especially in the male, enable a solifugid to pounce on its prey.*

it holds the flesh in its pedipalps and passes it to and fro between the jaws, which chew it to a pulp. Later, the solifugid sucks up the juices. It only occasionally drinks, scooping up water with the lower jaw mandibles. So long as there is food it will gorge itself until it can hardly move.

## Fainting females

Males are paler in color, smaller and more active than females. They have longer legs, narrower heads and smaller jaws. Some reports have claimed that males are more likely to flee from intruders, while females stand their ground and fight, but this has seldom been observed. In courtship the male taps the female's body with his pedipalps and strokes her until she is in a trancelike state, when he can approach without danger of being attacked. The male then opens her genital orifice with one of the chelicerae and discharges a drop of sperm, which he catches with the other chelicera and places inside the female. After this he scuttles away before she comes to. When ready to lay her eggs, the female goes into a deep burrow. She stays beside the eggs until they hatch. She then goes out to hunt for food for her offspring, which she feeds until they are old enough to leave the burrow.

## Medicine and myth

In some parts of their range there is a belief that solifugids are venomous. In fact, with the exception of the Indian species *Rhagodes nigracinctus*, which has venom glands, they can do little more than give a hard nip. Bites are normally reserved for their prey, but sometimes people are bitten; one species found in the North African desert is said to inflict a specially painful bite. J. L. Cloudsley-Thompson reported a belief in Egypt that solifugids "crawl in one's bed at night, bore into the crotch and lay their eggs." The people of Baku, on the Caspian Sea, believed the local solifugids to be most venomous when they first came out of hibernation in the spring. Their cure was to rub the wound with another solifugid— after first steeping it in boiling oil—to neutralize the venom. It has been suggested that the Hebrew word translated as "mouse" in the Old Testament of the Bible refers to a solifugid, and that the sores afflicting the Philistines were the results of their bites. He added that their rapid movements and hairy bodies gave an illusion of mice and that "they have been known to attack travelers asleep in the desert at night."

*A solifugid of the genus* Galeodes *devours a newborn rodent in North Africa. More common prey is insects and other invertebrates, but these arachnids eat almost anything they can overpower.*

---

## SOLIFUGID

| | |
|---|---|
| PHYLUM | **Arthropoda** |
| CLASS | **Arachnida** |
| ORDER | **Solifugae** |
| FAMILY | **Galeodidae** |
| GENUS AND SPECIES | ***Galeodes arabs*** |

ALTERNATIVE NAMES
**False spider; sun spider; wind scorpion; camel spider; solpugid**

LENGTH
**½–3 in. (1.2–7.5 cm)**

DISTINCTIVE FEATURES
**Huge, pincerlike chelicerae (jaws); male has flagellum (whiplike structure) on upper claw of chelicerae; 4 pairs of walking legs; sandy coloration**

DIET
**Mainly insects and other invertebrates; also lizards, small mammals, birds and their eggs**

BREEDING
**Age at first breeding: less than 1 year; breeding season: depends on rainfall; number of eggs: up to 200; hatching period: 2–10 days**

LIFE SPAN
**Up to 1 year**

HABITAT
**Deserts and other arid habitats**

DISTRIBUTION
**Throughout Africa**

STATUS
**Common**

# SONGBIRDS

THE ORDER PASSERIFORMES, commonly known as the passerines, is by far the largest order of birds. It contains about 5,300 species (60 percent of all birds), including some of the most widespread and successful species of all. It is a very diverse group of mostly arboreal (tree-living) species, although it also includes terrestrial and aerial species. Cosmopolitan in range, with the exception of Antarctica, its representatives live in habitats ranging from arid desert to rain forest, and from Arctic tundra to isolated oceanic islands.

The variation within the Passeriformes reflects the size of the group. The common raven, *Corvus corax*, is 25 inches (64 cm) long, has wings that may span 59 inches (150 cm) and weighs up to 3⁄10 pounds (1,560 g), more than 300 times heavier than tiny fellow passerines such as the pygmy sunbird, *Anthreptes platurus*, and the goldcrest, *Regulus regulus*. These diminutive species may weigh only ⅛ ounce (4–5 g), and the latter is only 3½ inches (9 cm) long.

## Classification

The passerines, or perching birds, are characterized by having four toes, three directed forward and one backward, all joining the foot at the same level. There are two suborders of passerines: the oscines and the suboscines. The majority of passerines have a highly developed vocal organ that enables them to sing beautiful and complex songs. These oscine

*Bluebirds (eastern bluebird, above) are related to American blackbirds and robins. They forage constantly, taking insects in the air, crawling insects, worms, larvae and fruits.*

passerines are known as the songbirds. The syrinx, the vocal organ of a bird, is only relatively poorly developed in most species, including the suboscine passerines. This group of about 1,200 species lives mostly in the Tropics and the New World. The group includes some very large families, notably the tyrant flycatchers, Tyrannidae, and the ovenbirds, Furnariidae. The oscines include more than 4,000 types of birds, in some 83 families. They have a more sophisticated syrinx than suboscines, and generally possess a correspondingly varied vocal repertoire. Consequently, it is within the oscines that birdsong reaches its most beautiful and complex heights.

| CLASSIFICATION | |
| --- | --- |
| **CLASS** Aves | |
| **ORDER** Passeriformes | |
| **SUBORDER** Oscines: songbirds | |
| **FAMILY** 83 | |
| **NUMBER OF SPECIES** More than 4,000 | |

The taxonomy of the Passeriformes is highly complex, as one would expect with so many families. Some very large, successful groups occupy distinct niches in different regions. The largest family is the Emberizidae, or buntings, sparrows and seed-eaters, with 321 species. This is the most successful family of seed-eating passerines, and straddles the Americas, Eurasia and Africa. The next three biggest families, the Sylviidae (Old World warblers), Muscicapidae (Old World flycatchers) and Timaliidae (babblers) are mainly insectivorous and are Eurasian and African. The largest American oscine family is the Thraupidae (tanagers), a group of 256 species of fruit-eaters and insect-eaters.

## Physical adaptations

The passerines are characterized by having feet suitable for gripping slender branches, twigs or grass stems. The hind toe is often stronger than the other toes, and is nonreversible. The greater potential for song that the oscines' better-developed syrinx provides them with is an adaptation for territorial defense and mate attraction.

Passerines generally lack the adaptations for swimming, wading or probing in deep mud that are shown by wildfowl or shorebirds. Most have short bills, reflecting their diet of seeds, fruits or insects. Some of the nectar-feeders, however, have longer bills, an adaptation for feeding inside flowers. The shrikes have powerful hooked bills for tearing apart small vertebrate prey. Passerines have short legs, an adaptation for perching rather than wading or swimming. As primarily terrestrial birds, they have unwebbed feet. This is true even of those species that have acquired some aquatic habits. The wings tend to be longer, relative to body length, in species that undertake long migrations, but no species have the very long wings characteristic of seabirds.

## Feeding techniques

Passerines fill most of Earth's ecological niches, and their feeding techniques reflect this. The bulk of species feed mostly on insects or seeds and other vegetable matter, or a combination of both. Some are generalists, while others are highly specialized. Just like humans, birds need an intake of proteins, carbohydrates, vitamins and minerals. Carbohydrates and fats are used primarily as sources of energy, but proteins are needed to construct new tissues. Proteins are therefore more important for reproduction, growth and molt. Fats and carbohydrates are crucial for migration. Birds that eat seeds for most of the year, such as buntings and many finches, supplement their diet with insects during the breeding season and may feed their young exclusively with this protein-rich diet.

Passerines may take insects from tree leaves or bark, on the ground, or on the wing. The Old World flycatchers habitually make sallies after flying insects from lookout posts, while other species may opportunistically attempt this strategy from time to time. Specialist aerial insectivores include the swallows and martins. The thrushes take a large proportion of fruit and berries as part of their diet.

The flowerpeckers and sunbirds consume a high proportion of flower nectar. Sunbirds have evolved a long, downward-curving bill that enables them to reach inside flowers to the nectar within. They also possess a long tongue, the tip of which is divided into three or four flaps that enable the birds to spoon up nectar. For most of its length, the tongue is rolled over on both sides, forming a tubular structure through which the sunbirds are able to suck the nectar up. Sunbirds probably play in important part in pollinating the flowers they visit. While feeding on nectar, pollen from the flower rubs off on the forehead. The birds then carry this pollen to the next flower they visit, rubbing it onto the stigma in the process of feeding. Plants such as tropical mistletoes are reliant on the actions of flowerpeckers, sunbirds and honeyeaters

*The scarlet tanager (male pictured) breeds in the eastern United States, migrating to the tropical areas of Central and South America for the winter.*

*The bobolink, D. orzyivorus, produces a sweet-sounding song. The male's black-and-white coloration becomes buffy with dark stripes, like that of the female, out of the breeding season.*

for pollination and might well die out without the birds' unwitting assistance.

Some of the insectivores are highly specialized feeders. For example, the oxpeckers are very differentiated insectivores, feeding primarily on ticks in the fur of large mammals, including buffalo, zebras and antelope, though they also take flies from time to time. Oxpeckers use their sharp, curved claws to secure a purchase on the skin of their host. The fairly long tail enables them to maintain their balance on near-vertical surfaces. Oxpeckers' bills are laterally flattened and the birds forage in the pelage or across the skin of the host using a scissoring action of the bill. Other super-specialists include the crossbills, which have specially twisted mandibles that can pry the seeds from pine cones, and the dippers that feed on invertebrates on the bottom of fast-flowing streams.

At the other dietary extreme, members of the crow family are practically omnivorous and can eat anything from human garbage, carrion and live vertebrates to berries and soft fruits. Many crow species exhibit predatory tendencies and frequently take the eggs and young of other species. The house crow, *Corvus splendens*, sometimes catches fish. Crows feed by holding the food with one foot while tearing at it with the bill.

Not all passerines eat food as they find it. Eurasian jays, *Garrulus glandarius*, hoard thousands of acorns in caches during the fall, revisiting these burial sites during the more barren months of winter. Shrikes store vertebrate kills in larders, such as thorn bushes or sharp barbed wire fences.

## Breeding strategies

There is marked sexual dimorphism (variation in appearance between the male and the female) in many oscines, the male having brighter plumage either just during the breeding season or year-round. The bright red plumage of the scarlet tanager, *Piranga olivacea*, is a prime example. By contrast, the female scarlet tanager has an olive-green plumage that blends in well with the forest foliage of the birds' summer habitat. This provides her with vital camouflage during the vulnerable period when she incubates her eggs and has young to rear. The ability of many male oscines to deliver complex songs is important in the breeding ecology of this suborder. Songs serve two roles: to advertize the availability of a territorial male to mate, and to warn off other males that may be potential challengers for territory or for mates.

Most passerines are monogamous, that is, one male mates with one female and the two form a pair-bond that may last for a breeding season, or for longer. For most, the pair-bond breaks up at the end of the breeding season, but that is not always the case. For example, the common raven and the jackdaw, *Corvus monedula*, (both members of the family Corvidae), pair for life. Monogamy presumably evolved in circumstances where the young had a better chance of surviving if both parents cooperated in rearing them. However, the time and energy devoted by monogamous males to rearing the young varies greatly from species to species. The male eastern bluebird, *Sialia sialis*, provides a site for the rearing of the young, but experimental removal of males has shown they are not essential for the purposes of rearing the chicks.

Other supposedly monogamous passerines are not always quite what they appear. Research has shown that the clutches of house sparrows (*Passer domesticus*) and reed buntings (*Emberiza schoeniclus*) with mixed parentage are not uncommon; sometimes they contain the offspring of more than one female, more than one male, or both.

Forty percent of the 122 well-studied European passerines adopt an alternative breeding strategy, that of polygyny, in which one male mates with more than one female, while each female mates with only one male. Polygynous species include the red-winged blackbird (*Agelaius phoenicius*), dickcissel (*Spiza americana*) and lark bunting (*Calamospiza melanocorys*). Polyandry, in which one female mates with more than one male, is not common among passerines.

Nests can be simple assemblages located on the ground, elaborate twig constructions or holes excavated in trees. Sometimes the abandoned nest of another species is adopted. Some species build a nest in a house or another artificial construction. Most species nest singly, but some prefer to congregate in loose colonies. This is true of the bearded parrotbill (*Panurus biarmicus*), which is also one of the most prolific passerine egg-layers: the female lays as many as 4 clutches of up to 11 eggs each in a season.

## A unique family: the shrikes

The 30 species of true shrikes in the family Laniidae are found throughout North America, Eurasia and Africa. In many respects they are not typical of the passerines. Thought to have evolved as forest birds, the true shrikes are now found in a variety of open woodland and forest habitats. Shrikes have a strong, sharp, hooked bill with a falconlike "tooth" on either side of the upper mandible near the tip, with a corresponding notch in the lower mandible. At one time, the bill shape led ornithologists to consider shrikes as relatives of the hawks. They are also rather raptorlike in behavior, and are capable of feeding on relatively large prey, including rodents and nestling birds, although they also take a wide range of insects, which they sometimes catch on the wing.

Shrikes characteristically scour the ground for prey from a vantage point. On seeing a suitable victim, they drop to the ground to catch it. Shrikes have strong legs and, like the crows, have the habit of grasping food with the foot. True shrikes are strong fliers, some species traveling thousands of miles a year on migrations.

True shrikes have hard and somewhat jarring calls that they use in order to contact other shrikes and as alarms. However, most species of true shrikes have distinctive songs and are accomplished mimics.

## Song

Birds do not produce sound in the same way that humans do. Rather than having a larynx at the top of the trachea, the vocal organ of a bird is a unique bony structure called the syrinx, situated at the lower end of the trachea. This voice box consists of bony rings with muscles and vibrating membranes. The volume and pitch of birdsong can be controlled by changing the air pressure passing from the lungs to the syrinx and by varying the pressure exerted by muscles on the syrinx's vibrating membranes. The oscines differ from other birds in having a more complex system of control, involving five to nine pairs of muscles, thus allowing greater variation.

Some passerines are able to sing very loudly. Others can mimic other birds. In fact, the the order of Passeriformes contains the most accomplished vocal mimics of all bird species. The marsh warbler, *Acrocephalus palustris*, for example, is known to mimic at least 99 European and 113 African species. The species with the greatest repertoire is the brown thrasher, *Toxostoma rufum*, which is estimated to sing more than 3,000 different mixes of song phrases.

## Migration

Many species of songbirds breeding in high latitudes are strongly migratory, particularly those that eat mainly insects. The onset of colder weather in the fall means that across much of North America and Eurasia the birds' main food source disappears. Billions of songbirds make the journey south every fall, returning once again the following spring in order to breed. New World wood warblers, thrushes and tanagers, to name only a few, desert their breeding grounds in the United States and Canada and head south to spend the winter months in Central and South America and the Caribbean. A similar exodus from Eurasia takes countless millions of Old World warblers, flycatchers, thrushes and chats to Africa and southern Asia. Some passerines may carry out a journey of 12,000 miles (20,000 km) or more during the course of one migratory round-trip.

*The sunbirds' long, decurved bills are specially adapted to help them feed on flower nectar and the insects that they find in flowers. This is a purple sunbird, Nectarinia asiatica.*

*Fieldfares, **Turdus pilaris**, feed on invertebrates such as insects and worms in summer, changing to a diet of seeds, fruits and berries in winter. They are a species of thrush.*

Long-distance migrants undergo a feeding frenzy before they set off and may add 35–50 percent to their body weight. This is vital because songbirds can lose 25–50 percent of their body weight in the process of migrating over large areas such as the Caribbean or the Sahara Desert. The sedge warbler (*Acrocephalus schoenobaenus*), common whitethroat (*Sylvia communis*) and pied flycatcher (*Ficedula hypoleuca*) are three Eurasian species that are thought to fly 40–60 hours nonstop for at least 1,320 miles (2,200 km) over the Sahara Desert. During its spring migration, the sedge warbler apparently flies nonstop 2,100 miles (3,500 km) from Uganda to the Middle East. Most smaller birds migrate at night when temperatures are lower, relative humidity is higher and there is less of a threat from predators. Not all migrants travel long distances, however. For example, some of the New World sparrows (family Emberizidae) migrate within North America. Species that do not need to migrate so far often travel by day.

## Relationship with humans

Songbirds have long featured in folklore and literature. The 16th-century English playwright William Shakespeare wrote about the singing qualities of a number of songbird species, notably the nightingale (*Luscinia megarhynchos*) and skylark (*Alauda arvensis*). Some species of passerines have long been kept as cage birds, both for their songs and for the beauty of their plumage. In the 19th century, large numbers were trapped for their colorful feathers, and although this practice is now rare, some species still suffer large casualties at the hands of humans for other reasons. In the Mediterranean region, some songbirds are considered delicacies, while others are shot for sport. The house sparrow and the house crow are two of a number of passerine species the lives of which have become inextricably linked with those of humans. They are nearly always seen close to human habitation and often build their nests inside artificial structures.

## Conservation

Habitat destruction is the greatest single threat to passerines. The breeding populations of many songbirds have been reduced by the use of pesticides and herbicides on farmland, the growth of cities and the clearing of forested areas. The threats are not always obvious: in North America, the break-up of large forests into smaller woods allowed the brood parasitic brown-headed cowbird, *Molothrus ater*, to lay its eggs in the nests of other passerines, reducing the number of young produced by the parasitized species in the process. Shade coffee plantations are an important habitat in Central and South America because they provide winter feeding for many North American wood warblers. However, these are being replaced by more modern methods of coffee cultivation. Some species of songbirds are now very rare and they will survive only if humans take steps to ensure their conservation. It is important to identify the places that are important for endangered songbirds, to discover where they breed, spend the winter and stop over on migration, and then to protect those places.

*For particular species see:*

# SONG SPARROW

*The most widespread species of sparrow in the United States, song sparrows can be found in almost any open, weedy or scrubby habitat.*

A LARGE NUMBER OF FINCHES living in North America are called sparrows. These include the grasshopper sparrow, *Ammodramus savannarum*, with its grasshopper-like buzz; the lark sparrow, *Chondestes grammacus*, of the fields; the chipping sparrow, *Spizella passerina*, which frequents gardens; and many others. At one time, the song sparrow was just one of many similar birds, and very little was known about its habits. However, in 1937, Margaret Morse Nice published the the first part of her work *Studies in the Life History of the Song Sparrow*, the results of an 8-year study of this familiar garden bird, and it is now one of the best-known birds in North America.

Song sparrows look rather like house sparrows, *Passer domesticus*, and are about the same size, but lack the house sparrows' black bib. Like many of their relatives, they have whitish breasts with brown streaks running from the chin to the belly. However, song sparrows are distinguished by the streaks converging to a dark spot just below the throat. The plumage changes with the habitat, varying from pale in desert areas to dark in humid regions. Song sparrows in Alaska are far larger than those found elsewhere.

The song sparrow ranges from Alaska and Canada, roughly on a line running through the southern shore of Hudson Bay, to the middle of the United States. It is more common in the east than in the drier west. In winter most song sparrows migrate southward to the southern United States and Central America. There is an isolated breeding population in central Mexico.

## Persistent songster

Scientists have divided the song sparrow into a number of subspecies, each of which lives in a particular habitat, such as salt marsh, desert or meadow. It is, however, best known as a garden bird with a lively song that can be heard for most of the year. Song sparrows may join in loose flocks during inclement weather but by the end of winter they take up their territories and singing may start as early as late January.

The male song sparrows sing until the end of the nesting season except, somewhat surprisingly, during the initial courtship. The main song period is in March, when the males have established their territories but before their mates have joined them. Even immature males sing; they have a soft warble, unlike the song of the adult, but this changes to the adult song as soon as they start to show territorial behavior.

Song sparrows eat mainly seeds and insects, but also a few berries, snails and spiders. In summer about half their food consists of insects.

## Large families

The outburst of singing in spring is a means of advertising to other male song sparrows that an area is already owned. When necessary, singing is reinforced by fighting. Courtship is just as violent as confrontation over territory. The aggressive male pounces on any female that enters his territory and collides with her. If she is seriously looking for a mate, the female withstands these assaults and the male accepts and courts her. The female then builds a nest, the male contributing only a few symbolic billfuls of nest material. The nest is made of dead grass and twigs and is lined with dried grass. Each pair of song sparrows may raise three broods of chicks, making a new nest for each. At the beginning of the nesting season, when the branches are bare, most nests are built on the ground, where they are concealed by herbage. Later nests are built above the ground among the foliage.

The usual clutch is three to five blue or grayish green eggs with brown or reddish brown spots. They are incubated by the female alone for

# SONG SPARROW

| | |
|---|---|
| CLASS | **Aves** |
| ORDER | **Passeriformes** |
| FAMILY | **Emberizidae** |
| GENUS AND SPECIES | ***Melospiza melodia*** |

**WEIGHT**
**⁷⁄₁₀ oz. (20 g)**

**LENGTH**
**Head to tail: 6⅓ in. (16 cm);
wingspan: 8¼ in. (21 cm)**

**DISTINCTIVE FEATURES**
**Small size; round head; stout bill; fairly
long tail; plumage varies according to
subspecies; coarsely streaked feathers;
2 bold brown throat stripes; central breast
spot; white chin**

**DIET**
**Seeds (all year); insects and other small
invertebrates (summer); also a few berries**

**BREEDING**
**Age at first breeding: 1 year; breeding
season: eggs laid February–August; number
of eggs: 3 to 5; incubation period: 12–13
days; fledging period: 10 days; breeding
interval: 2 or 3 broods per year**

**LIFE SPAN**
**Probably up to 12 years**

**HABITAT**
**Thickets of shrubs and trees among
grassland; forest margins; farmsteads;
sometimes parks and suburbs**

**DISTRIBUTION**
**Breeding: across U.S., apart from extreme
southeast; southern Canada north to
southern Alaska; small central Mexican
breeding population. Winter: southern part
of breeding range south to northern Mexico.**

**STATUS**
**Common**

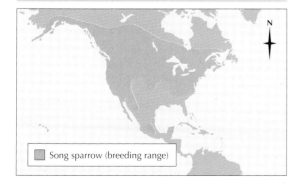

Song sparrow (breeding range)

12–13 days. Both parents feed the chicks, which leave the nest after 10 days. They continue to be fed for another 3 weeks, but the female gradually leaves them in the male's charge while she starts to build another nest for the next clutch of eggs.

## Parasitized by cowbirds

The song sparrow is one of the main hosts of the cuckoolike cowbird (discussed elsewhere). In Ohio, where Nice studied song sparrows, she found that 34 percent of the nests contained the eggs or nestlings of cowbirds. Sometimes, as many as 78 percent of the song sparrows' nests may be parasitized. The song sparrows are still able to rear their brood along with a cowbird chick, but on average they raise one less chick of their own.

## Alarm calls

Nice also investigated the reactions of song sparrows to different predators. She found that song sparrows have separate calls and postures to indicate alarm, fear and fright, and tested the reactions of captive song sparrows to both live predators and models. In the wild, cats arouse alarm but hand-raised song sparrows ignore them. Owls evoke fear in nature and model owls alarm captive adults. Hand-raised chicks ignore them until they are 3 weeks old and then show alarm. These and other observations suggest owls are recognized instinctively, whereas the birds have to learn that cats are dangerous.

It seems memory and experience play a very important part in the recognition of predators. Unpleasant experiences enhance fear; otherwise, reactions wane. This perhaps explains why young birds so easily fall victim to predators.

*Song sparrow nests are usually located at the base of shrubs or grass clumps. The nest is built by the female, and she is solely responsible for incubating the eggs.*

# SOUTH AMERICAN DEER

*In summer the marsh deer's coat is bright chestnut in color, changing to dark brownish red in winter.*

SOUTH AMERICAN DEER ARE small- to medium-sized deer, the smallest species standing only 16 inches (40 cm) at the shoulder. As is the case with a large number of South American animals, little is known of the 10 species of these extremely shy deer. They remain under cover for most of the day and consequently are not seen very often. The deer belong to five different genera, which have been grouped together here for convenience under the single heading of South American deer.

## Marsh deer

The marsh deer, *Blastocerus dichotomus*, sometimes called the swamp deer, is the largest of the South American deer. It has a head and body length of up to 6½ feet (2 m), stands up to 51 inches (1.3 m) at the shoulder and weighs up to 330 pounds (150 kg).

The deer's dark brown coat is coarse and long, its lower legs are black and the tail is yellowish red above and black below. The full-grown antlers are doubly forked, giving a total of four points. The hooves can be spread widely, an adaptation to walking on soft ground. The single species of swamp deer is found in Brazil, Paraguay and Uruguay.

Extremely shy and wary, swamp deer keep under cover by day and come into open spaces at night to feed on grass and water plants. They move about in groups of up to half a dozen in the high grass of wet savannas, along the damp edges of forests, along riverbanks and on wooded islands.

The deer readily take to water, both to feed and to escape. Little is known of their habits because they are so elusive. Zoologists believe the males do not drop their antlers at any set season. It is also reported that the fawns are born at various times of the year, suggesting that breeding may occur in any month. There usually is a single fawn, born after a gestation period of about 9 months.

## Deer among pampas grass

Another closely related species of South American deer also has no fixed breeding season in part of its range. This is the pampas deer,

# SOUTH AMERICAN DEER

CLASS **Mammalia**

ORDER **Artiodactyla**

FAMILY **Cervidae**

GENUS AND SPECIES **Marsh deer, *Blastocerus dichotomus* (detailed below); pampas deer, *Ozotoceros bezoarticus*; guemals, *Hippocamelus antisensis* and *H. bisulcus*; brocket deer, *Mazama americana*, *M. gouazoubira*, *M. rufina* and *M chunyi*; pudus, *Pudu mephistophiles* and *P. pudu***

ALTERNATIVE NAME
**Swamp deer**

WEIGHT
**Up to 330 lb. (150 kg)**

LENGTH
**Head and body: up to 6½ ft. (2 m); shoulder height: up to 51 in. (1.3 m)**

DISTINCTIVE FEATURES
**Dark brown coat; black muzzle; black lower legs; male antlers usually double forked; particularly long legs**

DIET
**Grasses, reeds and aquatic plants**

BREEDING
**Age at first breeding: probably 1–2 years; breeding season: October–February, perhaps year-round in some places; number of young: 1; gestation period: about 270 days; breeding interval: 1 year**

LIFE SPAN
**Up to about 20 years**

HABITAT
**Marsh, forest edges and other moist habitats**

DISTRIBUTION
**Brazil, Paraguay and Uruguay**

STATUS
**Endangered**

South American deer (all species)

*Ozotoceros bezoarticus*, slightly smaller than the swamp deer, with a reddish brown to yellowish gray coat, dark markings on the face and white underparts. The bucks are said to give out, from their foot glands, a strong, disagreeable odor that can be detected up to 1 mile (1.6 km) away. Pampas deer live among tall pampas grass, on the dry plains, in pairs or small herds. As the tall grass is cleared for cultivation, the deer are forced into the open and so become very wary. The range of the pampas deer goes farther south than that of the swamp deer, from Brazil to Patagonia. In the northerly parts of the range the pampas deer seems to have no fixed breeding season, but farther south, on the Argentine plains, it is said to breed at the end of summer.

## Deer in the Andes

The two species of guemals, or Andean deer, *Hippocamelus antisensis* and *H. bisulcus*, are stocky deer with short stout legs. The males have short antlers that branch once near the base, the front prong being the smaller of the two. About the same size as pampas deer, the guemal's coat is speckled yellow, gray and brown and there is a dark Y-shaped marking on the face. As in the Chinese water deer (discussed elsewhere), both sexes have long tusklike canines. However, these teeth do not hang low enough to be seen when the mouth is shut.

*The pampas deer, ranging from Brazil to Patagonia, lives in pairs or small herds in tall pampas grass on dry plains.*

The guemals are similar to pampas deer and swamp deer in habits except that they live at high altitudes in the Andes, in the forests or on grassy slopes, at 10,000–15,000 feet (3,000–4,500 m). They feed mainly on lichens and mosses and move down the mountainside into the valleys for the winter.

## Spike-horned deer

The most numerous species of South American deer are the brockets. The four species, *Mazama americana*, *M. gouazoubira*, *M. rufina* and *M. chunyi*, are distributed from the Guianas across Brazil to the Andes, as well as south to Paraguay and northward through Central America into Mexico.

Brockets are small animals, standing 2 feet (60 cm) or less at the shoulder and weighing up to 46 pounds (21 kg). They have bright red coats and short unbranched antlers that resemble the antlers of a red deer brocket, or first-year stag,

hence the English name for this South American deer. Brockets are characteristically high in the region of their hindquarters.

Brockets are solitary or go about in pairs, feeding in the early morning and again at dusk, on grasses and green shoots. They are extremely shy and secretive. Brockets are protectively colored and freeze when alarmed. Breeding takes place all year, with one spotted fawn at each birth.

## Smallest deer in the world

The two species of pudus are the smallest of all deer. The southern Andean pudu, *Pudu pudu*, which ranges from central Chile almost to the Straits of Magellan, grows up to 32 inches (81 cm) long with a tail measuring 1 inch (2.5 cm). It is up to 16 inches (40 cm) high at the shoulder and weighs 20 pounds (9 kg). The northern Andean pudu, *P. mephistophiles*, of Ecuador and Colombia, is smaller, lives at higher altitudes of 9,000–12,000 feet (2,700–3,600 m) and has a richer brown coat than *P. pudu*. Both species of pudus have short, spikelike antlers and narrow, pointed hooves. They are not often seen except when flushed out by hunting dogs. The fawns have three rows of white spots from the shoulder to the tail.

## Endangered and threatened

South American deer are preyed on, although not heavily, by anacondas, boas, jaguars and pumas. Local people hunt them for their flesh and hides and also because they damage crops, while the antlers are valued for their alleged medicinal purposes. The marsh deer, the pampas deer, the guemals, and one species of pudu deer are classified as endangered, and most of the other South American deer are classified as threatened or vulnerable.

Because little information is available about South American deer and they are so seldom photographed or kept in zoos, the impression has arisen that there can be few deer in South America. Although it is true that they do not form large herds, are probably less numerous in their populations and do not attain the sizes of deer elsewhere, South American deer are by no means insignificant. There are fewer than 50 species of deer in the world. Of these, nearly 20 species are from southern and Southeast Asia and 10 are from South America, a total greater than that for the number of species in Europe and North America put together.

*A young southern Andean pudu. Extremely shy in the wild, this species is the world's smallest deer.*

# SPADEFOOT TOAD

THE SPADEFOOT TOADS ARE named after the spadelike horny projection on the side of each hind foot with which they dig their burrows. The family of spadefoot toads is widely distributed over Europe, northwestern Asia and North America. They are usually 2–4 inches (5–10 cm) long with a soft skin that is moist like that of a common frog rather than dry and warty like that of a common toad. The color of the skin varies greatly between species. It may be gray, brown or green with red, white or black markings. There is also some variation in markings between the individuals of a single species.

The common spadefoot toad of Europe is found over much of Europe south of southern Sweden and extends into Asia as far as Iran. The best-known spadefoot toads live in North America. They are related to the European common spadefoot toad.

In Asia there live relatives of the spadefoot toads that are sometimes placed in the same family, but in a separate subfamily called the Megophryinae. Some of these Asian frogs species are called horned frogs, *Megophrys nasuta*. They have three unusual pointed projections of skin on the head, one on the snout and one above each eye.

## Spicy toads

Spadefoot toads are nocturnal, spending the day in burrows that they excavate by digging themselves in backward, pushing soil with their spadelike feet, while rotating their bodies. As they disappear beneath the surface, the entrance of the burrow caves in, concealing it. The burrowing and nocturnal habits of spadefoot toads make these animals hard to spot, even though they may be quite abundant. However, during the brief and sporadic breeding seasons, when the males can be heard calling, they are much more conspicuous. Spadefoot toads are mainly found in sandy areas where burrows are easy to dig. In dry weather they may burrow 6–7 feet (around 2 m) down to find moist soil.

Some spadefoot toads give a shrill cry when handled, which may be a means of deterring predators. They may also give off secretions from glands in the skin. In certain species, such as the Mexican spadefoot toad, *Scaphiopus multiplicatus*, small glands

give off an unpleasant-tasting secretion that also irritates the lining of the nose and mouth. The common spadefoot toad is called the *Knoblauchskrote*, or garlic toad, in Germany because of the strong smell of its skin secretions. The food of spadefoot toads is mainly insects and other small invertebrates. A few small lizards are also taken.

## Explosive breeding

While the European spadefoot toads are relatively unspecialized in their breeding behavior, among the North American spadefoot toads there are some very remarkable adaptations. These toads live in the dry parts of the southwestern United States and breed when shallow ponds are temporarily filled with rainwater. They therefore have to start breeding as soon as the ponds fill, and their offspring have to be independent before they dry up again.

Shortly after a storm the males search for water, and when they have found a suitable stretch of standing water, they start to call. Their calls attract other males, so a chorus builds up that eventually attracts the females, and pairs form for mating. The louder the chorus from any

*The eastern, or Hunter's spadefoot toad,* Scaphiopus holbrooki, *is a typical explosive breeder of North American arid habitats, meaning the breeding season is sudden, brought on instantly by seasonal heavy rains.*

*The common spadefoot toad of Europe does not possess the extreme arid-habitat adaptations of its North American relatives. Nevertheless, as a land-living amphibian, water conservation is always a concern.*

pond, the more females are attracted to it, which is an efficient way of ensuring rapid pairing. The eggs hatch in 2 days, a much shorter time than that known for any other frog or toad. The tadpoles grow very rapidly. The tadpoles of Couch's spadefoot toad are the fastest developers, completing development within eight days if conditions are optimal. Even so, sometimes the temporary pools dry up before they can change into toadlets. In some species the tadpoles gather in compact groups if the water level is dangerously low and wriggle together to form a depression in the mud where the remaining water can collect. This mud-stirring action can also expose food on the bottom.

The tadpoles also eat the bodies of other tadpoles that have died from starvation. This means that in bad conditions a few survive instead of all of them dying. It has also been found that when a pond is drying up, tadpoles that have fed on other tadpoles complete their development more rapidly, so increasing the chances of the strongest youngsters' survival.

## Each to its own

In the western United States there are four species of spadefoot toads that are very similar but only rarely interbreed. Where two or more species live in the same place, interbreeding is usually prevented by females responding only to the calls of males of their own species and by the slightly different behavior of different spadefoot toad species. For instance, Hunter's and Couch's spadefoot toads breed in the shallows, whereas the Plains, *S. bombifrons,* and Hammond's, *S. hammondi,* spadefoot toads prefer deeper water. There is also a great difference in the kinds of soil in which the American spadefoot toads prefer to live, and this also results in the species being kept apart. In Texas, Hunter's spadefoot toad

## SPADEFOOT TOADS

CLASS **Amphibia**

ORDER **Anura**

FAMILY **Pelobatidae**

GENUS *Pelobates, Scaphiopus,* **others**

SPECIES **European common spadefoot toad,** *P. fuscus;* **Couch's spadefoot toad,** *S. couchii;* **others**

LENGTH
**2–4 in. (5–10 cm)**

DISTINCTIVE FEATURES
**Flange, or spade, of hard skin on rear feet; eyes with vertical slit pupil**

DIET
**Mainly insects; a few small lizards**

BREEDING
*Scaphiopus:* **breeding season: after heavy rain; number of eggs: 10 to 500; hatching period: a few days; larval period: 2 weeks.** *Pelobates:* **slower development.**

LIFE SPAN
*S. couchii:* **up to 13 years**

HABITAT
*Scaphiopus:* **burrows in sandy or loose soils in arid areas; visits surface only in heavy rain, to breed**

DISTRIBUTION
*Pelobates:* **Europe and northwestern Asia.** *Scaphiopus:* **North America. Other genera (Megophryinae): China; Southeast Asia.**

STATUS
**Many species common**

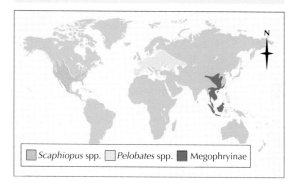

☐ *Scaphiopus* spp. ☐ *Pelobates* spp. ■ Megophryinae

likes sandy areas, whereas Couch's spadefoot toad prefers soil that is not sandy. This is a sufficient barrier to keep them apart, except where humans have disturbed the soil. At one place disturbed ground supports both species, and they interbreed occasionally.

# SPARROW HAWK

ANY HAWKS IN the genus *Accipiter* are called sparrow hawks. However, one of them, the sparrow hawk (*A. nisus*) of Europe, Africa and Asia, is much better known than the others. In fact it is one of the world's best-studied birds of prey. Many of the other sparrow hawks live in Southeast Asia, particularly on the numerous islands, or in Africa. They include the shikra (*A. badius*), Levant sparrow hawk (*A. brevipes*) and besra (*A. virgatus*). Also in the genus *Accipiter* are the goshawks, discussed elsewhere, and two North American species, the sharp-shinned hawk (*A. striatus*) and Cooper's hawk (*A. cooperii*).

The sparrow hawks are considerably different in appearance from the falcons. They have relatively short, rounded wings and fly with alternate bursts of rapid wingbeats and long glides. Sparrow hawks tend to hunt low over the ground, often in forest or other enclosed habitats, taking prey by surprise at the last moment. Falcons, by contrast, have long, pointed wings and are very fast-moving hunters that usually live in open habitats. In North America the American kestrel, *Falco sparverius*, which is a species of falcon, is sometimes confusingly called a sparrow hawk.

## Female much larger than male

Eurasian sparrow hawks, *Accipiter nisus*, vary in length from 11 to 15 inches (28–38 cm), the female being considerably larger than the male. The male is dark gray above with finely barred reddish brown underparts and a whitish chin. The female is darker and browner above but has whitish underparts with brown barring. Both have whitish napes, barred tails and long legs.

The sparrow hawk breeds in most of Europe except the treeless north, and in northwestern Africa. In Asia it lives in Iran and the Himalayas Mountains and east to Kamchatka and Japan.

## Victims of pesticide poisoning

Sparrow hawks live in woods and forests or in places where there are plenty of scattered trees. Even where abundant they are not seen as

*A female sparrow hawk yawns while standing guard on her treetop nest. There are four to six chicks in an average brood.*

## SPARROW HAWK

| | |
|---|---|
| CLASS | **Aves** |
| ORDER | **Falconiformes** |
| FAMILY | **Accipitridae** |
| GENUS AND SPECIES | ***Accipiter nisus*** |

**WEIGHT**
**Male: 4–7 oz. (115–200 g);
female: 6½–12 oz. (185–342 g)**

**LENGTH**
**Head to tail: 11–15 in. (28–38 cm);
wingspan: 21⅔–27½ in. (55–70 cm);
female up to 25 percent larger than male**

**DISTINCTIVE FEATURES**
**Strongly hooked bill; broad, rounded wings
with fingerlike primary feathers at tips; long
legs with powerful feet and strong talons;
long tail. Adult male: dark gray upperparts;
finely barred reddish brown underparts.
Adult female: dark brown upperparts; white
underparts with broad brown bars.**

**DIET**
**Almost entirely small and medium-sized birds**

**BREEDING**
**Age at first breeding: 2 years; breeding
season: eggs laid April–May; number of
eggs: usually 4 to 6; incubation period:
39–42 days; fledging period: 24–30 days;
breeding interval: 1 year**

**LIFE SPAN**
**Up to 12 years**

**HABITAT**
**Forest and woodland, including thickets,
orchards, parks and large gardens**

**DISTRIBUTION**
**Much of Europe, east through Asia to Japan;
northwestern Africa; Himalayas Mountains**

**STATUS**
**Common**

*Sparrow hawks are specialist bird-hunters. They usually prey on small birds such as sparrows and buntings, but can catch species as large as pigeons.*

frequently as falcons, because they spend most of their time in cover and typically hunt by flying behind hedges or through trees to take their prey unawares. Sparrow hawks are not, however, shy birds and have been known to fly into houses after their prey. Presumably the small birds, which are their main prey, take refuge in houses as a substitute for dashing into the cover of foliage, their usual place of safety.

The presence of sparrow hawks may be given away by the accumulation of pellets, feathers and bones of their prey around regular feeding posts. At one time sparrow hawks were very common in agricultural areas where there were sufficient trees to provide cover, but with the introduction of DDT and other long-lasting insecticides in the mid-20th century there was a serious decline in sparrow hawk numbers. The use of harmful pesticides has now been reduced and some of the most damaging have been largely abandoned. The sparrow hawk has recovered its numbers in many areas where formerly it was common.

Sparrow hawks that breed in northern regions migrate south in winter. North European sparrow hawks cross the Bosphorus and Gibraltar to winter in Africa.

### Sudden death to prey

The hunting habits of the sparrow hawk are as characteristic as those of the kestrels. It takes its prey by surprise, swooping on an unsuspecting bird and killing it with its long, sharp claws.

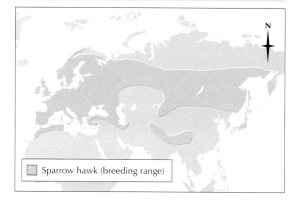

Sparrow hawk (breeding range)

Sometimes a sparrow hawk flips upside down and strikes its prey from underneath. Most prey is caught completely by surprise as the sparrow hawk glides rapidly along the edge of a wood or quarters a hedge by suddenly darting from one side to the other, but occasionally small birds are pursued through the air and are outpaced.

To anyone who sees a small bird suddenly disappear as a sparrow hawk flashes past, this method of hunting must seem breathtakingly efficient. It has been shown, however, that the sparrow hawk is successful in only 1 out of 10 attempts at a kill, and when it is successful the prey is often sick or already wounded.

The main prey of the sparrow hawk is small birds such as sparrows, buntings, finches, tits, warblers, robins and thrushes. Species up to the size of doves and pigeons are also caught, and female sparrow hawks tend to take larger prey than the males. Birds make up almost all of the sparrow hawks' diet, with the remainder consisting of a few insects and small mammals.

## Young birds practice hunting

The courtship of sparrow hawks is similar to that of other birds of prey, with both partners taking part in aerial chases and fights, the male some- times swooping at the female. Unlike most falcons, sparrow hawks build their own nests, constructing loose piles of interwoven twigs that are lined with green leaves. The female does most of the building but the male gathers much of the material and also feeds the female. He also feeds her while she incubates the clutch of four to six eggs. The chicks hatch out after an incuba- tion of 5½–6 weeks but they are brooded by the female for another week. She does not, however, help the male in hunting until the chicks are 3 weeks old. The chicks first fly when about 1 month old but cannot kill their own food for another week or more. They begin by catching insects, as more practice is needed before birds can be caught.

## Impact on prey species

At one time sparrow hawks were persecuted, like other birds of prey, because of damage done to game birds and poultry. The number of these birds that they do in fact kill is small, and it is now realized that sparrow hawks are useful. In one sample of sparrow hawk prey it was found that about 60 percent of the prey was detrimental to human interests. Many of the small birds that sparrow hawks kill are grain eaters.

*An adult male sparrow hawk bathing in a pool on the forest floor. Males are usually only three-quarters the size of females.*

# SPECTACLED BEAR

*The spectacled bear is native to tropical South American forests. Consequently, its fur is not as thick as that of bears from more temperate climates.*

THE SINGLE SPECIES OF spectacled bear, sometimes called the Andean bear, lives in tropical South America and is the only South American bear. It is one of the smaller bears, 47–71 inches (120–180 cm) long with a small tail of 2–2¾ inches (5–7 cm) and standing only about 27½–31½ inches (70–80 cm) at the shoulder. A full-grown male may weigh 220–374 pounds (100–170 kg). The shaggy coat is black or blackish brown with large circles or semicircles of white or buff around the eyes that have given the animal its common name. The muzzle is the same color and there is a white patch on the neck that extends in uneven streaks onto the chest. The head and chest markings vary widely in different animals and may be entirely lacking. The coat is much thinner than that of bears from temperate climates.

The spectacled bear lives in the forests, forest edge and grasslands in the foothills of the Andes, to an altitude of 13,200 feet (4,000 m) in western Venezuela, Colombia, Ecuador, Peru and western Bolivia and possibly in Panama. It usually keeps to the forested areas, but sometimes ranges into the higher clearings or into the plains, savannas and scrublands at very low altitudes.

## Strong for its size

The habits of the spectacled bear are still a subject of scientific debate. Like most bears, it is an expert climber and will climb trees to a height of as much as 80–100 feet (24–30 m) in search of food. It is strong for its size and is reputed to be able to snap saplings 3 inches (7.5 cm) in diameter. It is reported to make large nests of sticks in the trees, and although this has never been confirmed, it is quite possible, because other small bears do so, such as the sun bear, *Helarctos malayanus* (discussed elsewhere in this encyclopedia).

## Mostly vegetarian

Unlike most bears, the spectacled bear is thought to feed largely on leaves, fruits and roots, although in captivity some will take a certain amount of meat and there have been reports of wild bears preying on deer, guañacos and vicuñas. In Ecuador it feeds largely on the pambili palm, climbing the tree and tearing off branches, the leaves of which are eaten later on the ground. It also eats the buds of young palms, and tears open the green stalks to reach the tender pith inside. In northern Peru it feeds on the fruits of a species of *Capparis*, a genus of plants that includes *C. spinosa*, the unopened flower buds of which are harvested as capers. Meat forms less than 5 percent of the spectacled bear's diet.

## Breeding in zoos

The breeding habits of the spectacled bear are known only from the few animals that have been kept in zoos. Records dated 1951 concerning a well-established colony in the Zoological Gardens of Buenos Aires give the gestation period as 8–8½ months, the number of young as one or two per birth and the times of birth as June, July and September. The Zoological Gardens of Basel, Switzerland, reported that a pregnant female received by them from a dealer on November 25, 1952, gave birth to three cubs on February 17, 1953.

## Spectacled bears in captivity

The spectacled bear has never been numerous in zoos, probably because of the inaccessibility of its natural habitat. The Zoological Gardens of London obtained their first specimen in 1832. A spectacled bear that arrived at the New York Zoological Society in 1909 was thought to be the first to be seen alive in the United States. At first, the spectacled bear did not survive for long in captivity. Then zoologists realized that its diet was nearly

# SPECTACLED BEAR

| | |
|---|---|
| CLASS | **Mammalia** |
| ORDER | **Carnivora** |
| FAMILY | **Ursidae** |
| GENUS AND SPECIES | ***Tremarctos ornatus*** |

**ALTERNATIVE NAME**
**Andean bear**

**WEIGHT**
**220–374 lb. (100–170 kg)**

**LENGTH**
**Head and body: 47–71 in. (1.2–1.8 cm)**

**DISTINCTIVE FEATURES**
**Black or blackish brown, shaggy fur; variable whitish circles, semicircles or patches around eyes; whitish muzzle and patches on lower neck and chest; small tail**

**DIET**
**Fruits, tree sap and leaves of bromeliads (tropical American plants); very occasionally small rodents and birds**

**BREEDING**
**Age at first breeding: 2–3 years, possibly 4–5 years; breeding season: variable, birth peak in July, controlled by delayed implantation; number of young: 1 to 3; gestation period: 195–255 days; breeding interval: up to 2 litters per year**

**LIFE SPAN**
**Up to 38 years in captivity**

**HABITAT**
**Thorn forest, humid forest and grassland, from 1,650–13,200 ft. (500–4,000 m); plains, savanna, scrubland at low altitudes**

**DISTRIBUTION**
**Montane regions of western South America**

**STATUS**
**Vulnerable. Locally common in some areas; rare in most places due to hunting.**

Spectacled bear

*This female spectacled bear was photographed at the La Planada Nature Reserve in Colombia. Knowledge of the species derives mostly from the study of captive specimens.*

wholly vegetarian. When the diet was changed, captive specimens survived for much longer. One large male lived for 16 years in the New York Zoological Park, from November 1940 until June 1957. It was kept in an outside den that was snug but unheated, and proved to be quite hardy. In this climate, which was much colder than that of its natural habitat, it showed no inclination to become dormant during the winter, no matter how bad the weather. Its daily diet comprised 1 quart (1.1 l) of reconstituted evaporated milk, 18 apples and 7 loaves of raisin bread. It refused meat, or even a mixture of chopped meat and dog meal, although a young spectacled bear at the zoo later accepted small quantities of both. The longest living captive spectacled bear was kept in the San Diego Zoological Gardens. It lived for just over 38 years.

## Facial variations

The spectacled bear is an excellent example of the fact that animals vary in their physical characteristics just as people do. This variation is more pronounced in some species than in others, but it is always present.

In addition to the white markings on the faces of spectacled bears, there usually is a small amount of white on the snout. In some of these bears the white rings may be so enlarged and the white on the snout so extensive that the face becomes a white mask with just a little black around the eyes. At the other extreme, all the white on the face may be lost and a bear may be left with only a white muzzle. There are many gradations between these two extremes, so in effect no two spectacled bears are facially alike.

# SPERM WHALE

*The sperm whale is easily recognizable by its squarish, blunt head, which takes up nearly a third of its total body length.*

THE SPERM WHALE, *Physeter macrocephalus*, has a torpedo-like shape. The body is thickest just behind the massive head, in the neck region, behind which it tapers backward to the tail flukes. From the side the head has a squarish appearance, but from the front it is rounded above with a wide groove on each side, and below this it curves inward on either side to the upper jaw. The lower jaw is small compared to the rest of the head. It is Y-shaped, being wide at the angle of the mouth and narrowing rapidly so that the two halves run together toward the front. There are 16 to 30 conical teeth on each side of the lower jaw, each up to 8 inches (20 cm) long, and when the mouth is shut, these fit into sockets in the upper jaw, which at the most has a few vestigial teeth. There is a single blowhole to the left on top of the head. The interior of the mouth and the tongue are pearly white. A sperm whale has no true dorsal fin. Instead, there is a series of half a dozen low humps or ridges along the rear third of the back.

The greatest length recorded for a sperm whale is 84 feet (25.6 m), but in general males grow up to 60 feet (18 m) long, while the females are little more than half this length. On average, the male sperm whale weighs about 33 tons (29.7 tonnes) although it may reach a maximum of about 50 tons (45 tonnes). The female weighs about 13 tons (11.7 tonnes), up to a maximum of 14 tons (12.6 tonnes). Sperm whales are dark gray on the back and lighter gray on the flanks. The underside is usually silvery gray to white.

Although most common in the warmer seas, sperm whales are found throughout all oceans, including both north and south polar seas. The main stock is, however, found between latitudes 40° North and 40° South.

The two species of pygmy sperm whales, genus *Kogia*, are up to 12 feet (3.6 m) long and lack the massive barrel-like head of their relatives, being a little more dolphin-shaped.

## Sleeping leviathans

Sperm whales move about in schools or herds of mainly three kinds: harems made up of cows and calves and usually led by a dominant male; bachelor schools; and loose groups of lone bulls. There are occasional rogue bulls that are particularly aggressive. The harems in particular keep within the latitudes of 40 degrees above and below the equator, which is why usually only bulls are found in temperate and polar seas. In general, all that is visible of sperm whales is the blow (exhalation), or at most their backs, except when they breach, leaping clear of the water and falling back flat on the surface. They surface for 10 minutes at a time, take 60 to 70 breaths and then submerge for up to an hour, diving almost vertically. Sperm whales are renowned for their

## SPERM WHALES

CLASS **Mammalia**

ORDER **Cetacea**

FAMILY (1) **Physeteridae**

GENUS AND SPECIES **Sperm whale,** *Physeter macrocephalus*

FAMILY (2) **Kogiidae**

GENUS AND SPECIES **Dwarf sperm whale,** *Kogia simus*; **pygmy sperm whale,** *K. breviceps*

WEIGHT
*P. macrocephalus*: **22–55 tons (20–50 tonnes)**

LENGTH
*P. macrocephalus*: **36–59 ft. (11–18 m)**

DISTINCTIVE FEATURES
**Large, squarish heads.** *K. breviceps* **and** *K. simus*: **blue-gray body; false gills (lines where gills would be in fish); small hooked dorsal fins.** *P. macrocephalus*: **grayer color; wrinkled skin; rounded hump instead of dorsal fin; head about ⅓ of total length.**

DIET
**Mainly squid, fish and crustaceans**

BREEDING
*P. macrocephalus.* **Age at first breeding: 7–13 years (female), 25–27 years (male); breeding season: peaks in spring; number of young: 1; gestation period: about 420–570 days**

LIFE SPAN
**Up to about 75 years**

HABITAT
**Oceanic waters; occasionally near coasts**

DISTRIBUTION
**Virtually worldwide**

STATUS
*P. macrocephalus*: **endangered.** *K. simus* **and** *K. breviceps*: **not known.**

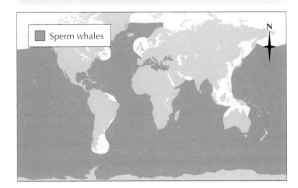

Sperm whales

N

deep dives, during which they are said to emit sounds for the purposes of assessing water depth and the nature of their surroundings. Cable ships have often hoisted broken submarine cables to the surface for repairs only to find a sperm whale tangled with the cable, usually by its lower jaw.

Sometimes a sperm whale performs an action known as a lobtail, in which it stands on its head with the tail flukes 30 feet (9 m) above the surface. It may then forcefully smack the surface with its tail, a sound audible for miles around.

Over the centuries, seagoing vessels ranging from merchant vessels to warships have collided with sperm whales. During World War II a United States destroyer experienced a heavy jolt one night and rapidly began to lose speed. The ship's company took to their boats, thinking they had been torpedoed. Next morning a sperm whale was found transfixed to the bows. There are a number of similar records. The probable cause of such accidents is that the ships hit sleeping sperm whales.

Sperm whales have been reported playing with floating planks. This is interpreted as play that is related to the instinct of these whales to rescue a calf, for although a sperm whale frequently deserts an injured mate, it apparently responds to the distress cries of a nursing calf. There are reports of an adult taking a calf in its mouth and rising to the surface with it.

## Meals of giant squid

The main food of sperm whales is squid and cuttlefish, including giant squid, and scars on the whales' bodies, apart from those resulting from fighting between males, are often caused by the hooks and suckers of their prey. Some of the sucker marks may be 4 inches (10 cm) across, and suggest encounters with extremely large squid.

*Sperm whales are named after the oil-like spermaceti housed in an organ above their upper jaw. This substance may help the whale to adjust its buoyancy when diving.*

*Sperm whales often dive to depths of 2,640 feet (800 m) to hunt, but may descend to 1¼ miles (3 km). In tropical seas, this could mean a temperature drop from 77° F (25° C) to nearly the freezing point.*

There is a record of a squid 34 feet 5 inches (10.3 m) long and weighing 405 pounds (890 kg) found in the stomach of a 47-foot (14-m) sperm whale harpooned off the Azores in 1955. Harpooned sperm whales sometimes regurgitate the remains of squid, and one was seen to give up 75 to 100 squid in this way, most of them being 3–4 feet (0.9–1.2 m) long, the usual size of squid they take. They also take seals and fish; one was found to have a 10-foot (3-m) shark in its stomach.

## One calf every four years

Puberty occurs in the male when he is about 40 feet (12 m) long, but he is not sexually mature until he reaches a length of about 46 feet (13.8 m). The female matures sexually at a length of about 28 feet (8.4 m) and her reproductive cycle lasts slightly more than 4 years. This comprises a pregnancy of 14½ months, 24–25 months suckling a calf (there is usually only one at a birth) and a 9-month resting period before breeding again. Most of the information about birth in sperm whales comes from South Africa. There is a pairing season off Durban on the east coast from October to April, with a peak in December.

After mating, the females go north to spend the winter in equatorial waters, returning south in spring. The young males go with them, the mature bulls following later. The newly born offspring is 13–14 feet long (3.9–4.2 m).

## Spermaceti and ambergris

The sperm whale fishery began in the early 18th century and was mainly controlled by North American whalers from the New England coast. At its height, the sperm whale industry produced 4 million gallons (18 million l) of oil from the blubber, which may be up to 14 inches (35 cm) thick. The fishery was over by the end of the 19th century, partly because the stocks in the western Atlantic were seriously reduced, but mostly because whale oil was replaced by mineral oil for lighting. There are still fisheries for sperm whales off the coasts of the Azores and South Africa, in the Pacific and, to a lesser extent, elsewhere.

The large reservoir of spermaceti oil in the sperm whale's head was regarded as one of the more commercially valuable parts of its body. Although its purpose remains a matter of scientific debate, it is possible that the spermaceti contracts as the whale dives into cooler waters, helping the whale to reduce its buoyancy. This would enable the animal to descend to greater depths, where it hunts the majority of its prey, more effectively. As many as 15 barrels of spermaceti could be tapped from one whale. This oil solidifies to a soft yellow wax and was used to make candles, cosmetics and medicines, and for other purposes, such as waterproofing fabrics.

The substance the sperm whale is best known for, however, is its ambergris. This is gray to black in color, lighter than water and has an offensive smell when it is fresh from the whale's intestine, though it later develops a characteristic sweet, earthy odor. Ambergris usually is found in small pieces floating on the sea. Although it is only a by-product of the whale's digestion, ambergris was once valued highly for medicinal purposes and as a base for the finest perfumes. The process of its production remains uncertain, but it almost certainly has something to do with the squid and cuttlefish eaten by the whale.

# SPIDER CRAB

SPIDER CRABS ARE CHARACTERIZED by their relatively long, spindly limbs and by a carapace (shell) that is longer than it is wide. There are many species, varying dramatically in size. The largest on record, measuring 12½ feet (3.8 m) across its outspread limbs, was a specimen of *Macrocheira kaempferi*, the giant crab of Japan. The body measured a mere 15 inches (40 cm) across. Its weight is not known, but a crab spanning just over 12 feet (3.7 m) weighed 41 pounds (18.5 kg). Another large species, *Leptomithrax spinulosus*, reaches 2½ feet (75 cm) across the extended claws and 6 inches (15 cm) across the back. It lives off southeastern Australia.

Many spider crabs are very small—like the long-beaked spider crab of Europe, *Macropodia rostrata*, with a shell less than ½ inch (12 mm) across. Commonly found under rocks and among weeds low on the shore, it is the most spiderlike European species, with its long, slender, bristly legs. The first of the four pairs of walking legs are the longest, being at least three times the length of the body. Its claws are shorter than its walking legs and are rather heavy in the male. The triangular or pear-shaped body, reddish in color, narrows to the front, where it ends in a pair of prongs between the stalked eyes. This two-pointed beak, or rostrum, is typical of spider crabs. The body surface is variously spined or covered in warts, depending on the species. The abdomen, or tail, tucked beneath as in other crabs, may be six- or seven-jointed.

The largest European species is the spiny spider crab or thornback crab, *Maia squinado*, whose body may grow to be 2–7 inches (5–17.5 cm) across. It can be found on the shore but prefers depths of 90–600 feet (30–200 m), and is very abundant in some years, to the displeasure of the lobster fishermen whose pots it infests. Its legs are slender but not especially long. The body is reddish with pink, brown or yellowish markings. About six hard spines project on each side of the shell, and others cover the surface.

## Usually lethargic

Spider crabs tend to be slow-moving carrion scavengers, relying more on concealment than on speed for protection. The sluggishness of these and some other crabs is linked by scientists to a high level of magnesium in their blood in comparison with that in more active crustaceans; magnesium salts are well known for their anesthetic action. There is a tendency for dissolved substances in the body fluids of marine animals to be lost by osmosis to the seawater when there are no physiological mechanisms to prevent it. The concentration of magnesium salts in spider crab blood is close to that in seawater. It seems that the high levels of magnesium in these species reflect an inability to keep them lower—in contrast with, for example, the shore crab, which can osmoregulate to some extent. This explains the inability of spider crabs to penetrate the brackish water of estuaries.

The kelp crab, *Pugettia producta*, is more active than most spider crabs and clings strongly to one's fingers if picked up. It is found on seaweedy beaches on the Pacific Coast from British Columbia to Baja California. It does not purposely adorn its carapace, but barnacles and anemones are often found clinging to it.

## Four stages of crab

The life history of a spider crab is similar to that of other crabs, involving two zoea stages followed by a transformation into the more crab-like megalopa. The zoea larva has large eyes, an elongated abdomen and two large spines, one forming a downwardly directed beak and the other projecting upward from the back. Molting ceases at puberty.

## Long legs, green fingers

Some spider crabs stay on the background they match, such as *Parthenope investigatoris*, a spider crab of the Indian Ocean, which looks like a piece

*An arrow crab, Stenorhynchus seticornis, perching on a coral head in Cuban waters. This delicate little spider crab hides under ledges by day.*

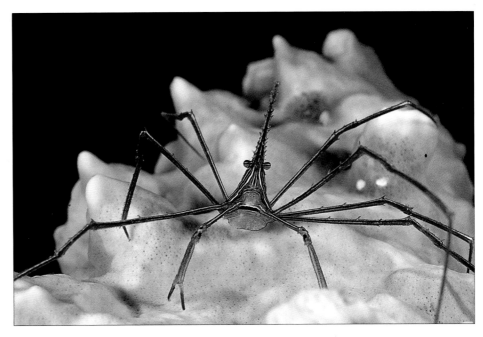

## LONG-BEAKED SPIDER CRAB

| | |
|---|---|
| PHYLUM | **Arthropoda** |
| CLASS | **Crustacea** |
| ORDER | **Decapoda** |
| FAMILY | **Majidae** |
| GENUS AND SPECIES | ***Macropodia rostrata*** |

ALTERNATIVE NAMES
**Slender-legged spider crab; long-legged spider crab**

LENGTH
**Shell length: 1¼ in. (28 mm); shell width: ⅔ in. (16 mm); leg length: 2½–2¾ in. (6–7 cm)**

DISTINCTIVE FEATURES
**Pear-shaped body narrowing at front; stalked eyes; slender, hairy second to fifth legs; bristly brown shell, with gray, yellow or red tinge, usually camouflaged with organic matter; female claws slender; male claws stouter and longer**

DIET
**Organic seabed debris, including carrion**

BREEDING
**Female releases eggs between May and January; eggs hatch into zoea larvae, developing into megalopa larvae; undergo several molts before changing into crabs**

LIFE SPAN
**Approximately 3–5 years**

HABITAT
**Under rocks and among weeds on seashores up to depths of 100 yd. (90 m)**

DISTRIBUTION
**Atlantic coasts, from Norway to West Africa, Azores and Mediterranean Sea**

STATUS
**Common**

*Spider crabs can swim as well as scuttle in typically sideway crab fashion. Despite their fearsome appearance, spider crabs are harmless to humans, feeding on carrion and other organic matter on the seabed.*

of the worn coral among which it lives. Others change color to suit the background on which they happen to be. A number of species disguise themselves with seaweeds, sponges or hydroids, each piece plucked and fastened onto the body. A sticky secretion from the mouth may be used to attach the fragments, but in many spider crabs the surface of the shell bears special hooked or serrated bristles or bundles of hairs that curl like tendrils around whatever is planted on them.

Many of the smaller, more common spider crabs cover their bodies in such a fashion, and *Camposcia retusa* of tropical reefs is called the harlequin crab for the assortment of gaily colored fragments it places on its shell. By contrast, *Inachus dorsettensis*, a small yellowish brown crab of British waters, camouflages only its first pair of walking legs. In the United States spider crabs are known, appropriately, as decorator crabs.

Camouflaging species tend to occur more often in clear water than where visibility is poor. The mobile garden is naturally replaced each time the shell is molted, but a new one may be replanted in special circumstances. When some species of the genus *Hyas* were put, covered with seaweed, in an aquarium with sponges and no weed, they removed their inappropriate camouflage of weed and replaced it with sponge. When scientists removed all covering, the crabs became very perturbed and uncharacteristically active.

*Podochela hemphilli*, a spider crab on the California coast, clothes itself with the more delicate red seaweeds, and while it is stationary it waves the weed back and forth by rocking its body, so adding to the realism. When hungry, it eats pieces of its camouflage.

The large European *Maia squinado* decorates its shell when young, but gives up the habit as it grows larger and better able to defend itself.

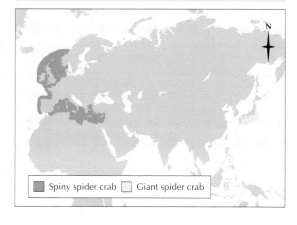

Spiny spider crab ☐ Giant spider crab

# SPIDER MONKEY

SPIDER MONKEYS ARE among the best known and most highly specialized South American monkeys. Like other monkeys, their noses have sideways-facing nostrils, divided by a very broad septum (dividing membrane). They also have three premolars in each half of each jaw, unlike African and Asian monkeys, which have only two premolars. Their skull characters and anatomical details also make South American monkeys less like humans than the Old World monkeys and apes, although in humans the third molars, or wisdom teeth, are often missing.

Spider monkeys are slenderly built but potbellied. The fur usually is rather wiry and sparse on the underside. The tail is prehensile (adapted for seizing or grasping) and can wrap around and cling onto branches. On the underside of its tip, the tail has a completely hairless area a few inches long, marked by wrinkles and ridges that resemble fingerprints. These ridges enhance the tail's ability to grip. The hands are modified as hooks, with long, narrow palms, long curved fingers and much-reduced thumbs.

There are two genera and seven species of spider monkeys. The common spider monkey, *Ateles paniscus*, has coarse, wiry hair and may be a variety of colors, according to race. One race may be black and buff, another wholly black and yet another black with a red face and genitalia. The woolly spider monkey, from the dry hardwood forests of southeastern Brazil, has thicker, more woolly hair and is more robustly built. It is yellowish or grayish brown and rather darker on the head and neck. It resembles a woolly monkey, genus *Lagothrix*, but the resemblances are thought to be only superficial. The woolly spider monkey, *Brachyteles arachnoides*, is heavier than the common spider monkey, being about 21 pounds (9.5 kg) as against 12–15 pounds (5.5–7 kg), but both species have a head and body length of about 15–25 inches (38–64 cm), with a tail length of 20–35 inches (50–90 cm). Woolly spider monkeys are rare today because their forest homes are being cut down for land cultivation. They are also hunted for their meat.

## Apelike habits

Spider monkeys are very versatile. On the ground or along a branch, they walk on all fours with the tail curled into an S-shape and held

over the back. On the ground the fingers are often held bent, so that they walk partly on their knuckles. Spider monkeys often walk upright with the arms held either down by their sides or grasping a nearby branch or rail. In the trees, they usually move about using the two hooklike hands and the prehensile tail. They commonly swing along by their hands in the manner of a gibbon, with the tail occasionally giving additional support. From time to time they may drop 20–30 feet (6–9 m) through the trees at once.

Spider monkeys live high up in the trees, only occasionally descending to the ground. They live in troops of varying size, which are constantly splitting and reforming. The troops comprise groups of females with their young, and may or may not be accompanied by an adult male. Males and females are of equal ranking and females have a weakly marked rank order among themselves. Males groom females and tend to be intolerant of each other in the presence of females. High-ranking males groom themselves more often than they are groomed by others.

## Versatile tails

Spider monkeys regularly use their tails to act as useful extra limbs. When they sleep, they sit huddled, often two or three together, with the tail holding onto a support. Its use seems to be automatic, winding around any object with

*A common spider monkey,* **Ateles paniscus.** *Once thought to include only fruits, the diet of spider monkeys is now known to feature nuts and invertebrates.*

which it comes into contact. In a zoo, a spider monkey regularly holds a peanut in its tail in order to transfer it to its mouth.

Spider monkeys eat fruits, nuts, invertebrates, birds' eggs and leaves. They are selective about their food, sniffing a fruit to see if it is ripe, and biting then rejecting it if it is unsatisfactory. Spider monkeys may taste and even half-eat a great deal of food before discarding it.

## Grappling behavior

Spider monkeys perform a pre-mating behavior called grappling in which the male and female sit opposite each other on a branch, push and pull, and cuff each other with their fists, and slap and bite one another. Afterward, the male may chase the female, often roaring as he does so. Then they become quiet and return to their sitting positions, with the female generally sitting on the male's lap, facing him. This behavior seems to occur mostly in the evening. Mating probably takes place at night because scientists have not yet observed it, at least in the wild. Young are born year-round, but there may be birth peaks, especially for woolly spider monkeys, at times when food is particularly abundant.

The female spider monkey's sexual cycle lasts 24–27 days and gestation is of 200–230 days' duration. The single young clings to the mother's belly at first, but after about 4 months it transfers to her back, with its tail twisted around hers. The young are very playful. They are black for the

*The rain forest canopy of Amazonia, Brazil, provides spider monkeys with the perfect environment in which to live.*

## SPIDER MONKEYS

| | |
|---|---|
| CLASS | **Mammalia** |
| ORDER | **Primates** |
| FAMILY | **Cebidae** |
| GENUS | *Ateles* and *Brachyteles* |
| SPECIES | *Ateles*: 6 species, including common spider monkey, *A. paniscus*. *Brachyteles*: woolly spider monkey, *B. arachnoides*. |

WEIGHT
*Ateles* species: 13–22 lb. (6–10 kg);
*B. arachnoides*: up to 26½ lb. (12 kg)

LENGTH
Head and body: 15–25 in. (38–64 cm);
tail: 20–35 in. (50–90 cm)

DISTINCTIVE FEATURES
*Ateles* species: long prehensile tail; red, gray, brown or black fur on back; sparsely furred lower surfaces; bronze or golden flanks; white eye rings in some subspecies. *B. arachnoides*: much denser fur; bright red, naked face.

DIET
Fruits, nuts, invertebrates, eggs and leaves

BREEDING
Age at first breeding: *Ateles* species: 5 years (male), 4 years (female); *B. arachnoides*: 7 years. Breeding season: *Ateles* species: all year; *B. arachnoides*: May–September. Number of young: 1. Gestation period: *Ateles* species: 200–230 days; *B. arachnoides*: 240 days. Breeding interval: 36 months.

LIFE SPAN
Up to 48 years in captivity

HABITAT
Canopy of tropical forest

DISTRIBUTION
Panama south to surrounds of Amazon Basin

STATUS
All species endangered or vulnerable

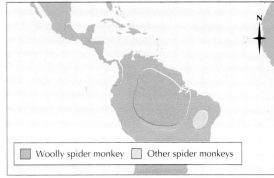

Woolly spider monkey   Other spider monkeys

first 6 months or so and then assume adult coloration. Spider monkeys can live for more than 20 years in the wild.

## Predators and lost habitat

From time to time, birds of prey, jaguars and small cats undoubtedly take adult spider monkeys or, more likely, their young. Humans also are predators. The native peoples of South America have been hunting spider monkeys for food with blowpipes for several thousand years without seriously decreasing the monkey populations. The real threat to these monkeys, as to all South America's fauna and even to the native peoples themselves, is considered to be the attitudes of those organizations that own the forests. Timber and rubber concessions have led to deforestation, which has resulted in the loss of habitats and, consequently, of wildlife.

## South American "apes"

The Omomyidae, the family that gave rise to the higher primates (monkeys, apes and humans), lived in northern Eurasia and North America about 40 million to 60 million years ago. The more familiar monkeys, such as the baboons, macaques, langurs and colobus monkeys, and the apes, from which humans evolved, lived in the Old World. These monkeys are called Catarrhines, because of their narrow noses with forward-facing nostrils.

The Omomyids may have migrated to South America across a land bridge, or on vegetation rafts, or by spreading from island to island. There they evolved into the New World monkeys, which are called Platyrrhines because of their wide noses with sideways-facing nostrils.

South America had similar environments to those of the Old World, but they were exploited by different types of monkeys. Over time, these came to resemble the Old World monkeys and apes, although their common ancestors were the Omomyidae, not monkeys. In Asia and Africa, similar habitats are occupied by guenons, mangabeys, mandrills, colobus and langurs, all leading slightly different lifestyles and so avoiding competition with each other. This form of noncompetitive coexistence also applies to the apes.

In South America, spider monkeys exhibit a number of apelike characteristics. For example, their arm bones are strong, with large muscles,

and their arms are longer than their legs. The hands are long and narrow, and the fingers are long and curved, giving the hands a hooklike shape. Apes have short thumbs and those of spider monkeys are also much reduced. Through specializing as fruit eaters, like the apes, spider monkeys too have become brachiators, swinging from the trees, using one arm and then the other, as gibbons do. The fruit they eat generally grows out along the branches, which are too thin to walk along. As a result, the best way for the monkeys to gather it is to hang by the arms below the branches.

*Spider monkeys have long, muscular arms, hooklike hands and a strong prehensile tail, key adaptations for an arboreal (tree-dwelling) way of life.*

# SPIDERS

*The eyes of a killer:* **Marpissa muscosa,** *like many jumping spiders, uses keen vision to judge its leaps upon prey.*

ONE OF THE MOST DIVERSE and successful of all arthropod groups, the spiders flourish in a wide range of habitats, from rain forests to deserts, mountain tops and tundra to salt marshes and ponds. There are even some species that manage to survive in turbulent marine rock pools.

Spiders are arachnids and are allied with groups such as the scorpions (Scorpiones), whip scorpions (Uropygi) and mites (Acarina). They are an ancient group, with the earliest indisputable fossils dating from around 380 million years ago.

## Classification

The spider order, Araneae, comprises three infraorders. The smallest contains the liphistiid burrowing spiders. These form a distinct sister group to all other spiders and have a number of unusual characteristics, including plates on the abdomen that recall the segmentation found in the earliest known spiders. These spiders are limited to a single family, found in southern Asia. A second group, the Mygalomorphae, is more widespread and includes several familiar spiders, including the funnel-webs (Dipluridae), the tarantulas (Therophosidae) and the trapdoor spiders (Ctenizidae). However, by far the largest number of spiders are classified within the third infraorder, the Araneomorphae, which branched off from the mygalomorphs around 250 million years ago. Among their number are the well-known orb-web (Araneidae), jumping (Salticidae) and crab (Thomisidae) spiders.

Araneomorphs and mygalomorphs can very loosely be distinguished by sight, in that the latter tend to be of the large, hairy-bodied and hairy-legged type. A better distinction is drawn from the structure of the fangs. Araneomorph fangs are arranged so that they close from the sides, like pincers. These opposable fangs allow the spider to bite its prey with force. Mygalomorphs are unable to do the same as their fangs point downward from their attachment to the chelicerae, with the result that they can stab prey but not bite it. Mygalomorphs must generate momentum for the stab by raising the body, extending the fangs, and then punching down and forward. Araneomorphs need only use the fangs to deliver a fatal bite.

| CLASSIFICATION |
| --- |
| **CLASS** |
| Arachnida |
| **ORDER** |
| Araneae |
| **SUBORDER** |
| Araneomorphae: true spiders; |
| Mygalomorphae: mygalomorph spiders; |
| Liphistiomorphae: Asian burrowing spiders |
| **NUMBER OF SPECIES** |
| More than 30,000 |

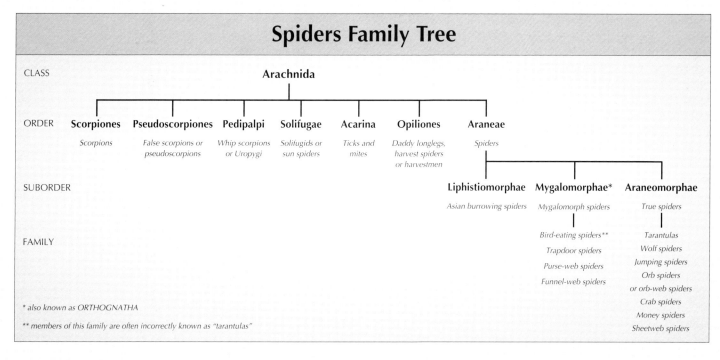

# Spiders Family Tree

| CLASS | | | | Arachnida | | | | | | |
|---|---|---|---|---|---|---|---|---|---|---|

| ORDER | **Scorpiones** | **Pseudoscorpiones** | **Pedipalpi** | **Solifugae** | **Acarina** | **Opiliones** | **Araneae** |
|---|---|---|---|---|---|---|---|
| | *Scorpions* | *False scorpions or pseudoscorpions* | *Whip scorpions or Uropygi* | *Solifugids or sun spiders* | *Ticks and mites* | *Daddy longlegs, harvest spiders or harvestmen* | *Spiders* |

| SUBORDER | | **Liphistiomorphae** | **Mygalomorphae\*** | **Araneomorphae** |
|---|---|---|---|---|
| | | *Asian burrowing spiders* | *Mygalomorph spiders* | *True spiders* |

| FAMILY | | | *Bird-eating spiders\*\** | *Tarantulas* |
|---|---|---|---|---|
| | | | *Trapdoor spiders* | *Wolf spiders* |
| | | | *Purse-web spiders* | *Jumping spiders* |
| | | | *Funnel-web spiders* | *Orb spiders or orb-web spiders* |
| | | | | *Crab spiders* |
| | | | | *Money spiders* |
| | | | | *Sheetweb spiders* |

*\* also known as ORTHOGNATHA*

*\*\* members of this family are often incorrectly known as "tarantulas"*

## Physical adaptations

The spider body is divided into two sections: the cephalothorax, which combines the head and midbody, and the abdomen. A narrow waist joins these sections. In addition to the eight legs, each tipped with two or three claws, there are a number of other appendages. These include the palps, which provide sensory information and also play a major role in mating; the chelicerae, which act as jaws and to which are attached the fangs; and the spinnerets, which yield silk.

Many spiders have one or two pairs of book lungs in the abdomen; these consist of a series of layers filled with hemolymph (blood), which flows past tiny air pockets to extract oxygen and ferry it to the heart. Others have tracheal systems, similar to those of insects. The hemolymph itself contains an oxygen-carrying molecule, hemocyanin, which tends to add a pale blue color to the hemolymph.

Spiders have up to four pairs of eyes. Day-active species that hunt without the use of webs, such as lynx (Oxyopidae) and jumping spiders, have good eyesight, with the main eyes at the center providing high definition and the smaller, lateral eyes offering a wide field of view. Other spiders are more reliant on touch and taste, both of which are detected through sensory or tactile hairs on the legs and palps. Special hairs known as trichobothria can detect air movement, such as that caused by an insect's flapping wings.

## Getting around

Spiders walk in an unusual way; the legs are flexed using muscles, but to extend them the spider increases the pressure of the hemolymph inside the body, using the hydraulic properties of the fluid. In animals without wings, dispersal over wider areas may be problematic. Spiders overcome this by ballooning, especially when they are tiny spiderlings. In mygalomorphs this involves simply climbing to a high point and jumping off into the wind. Araneomorphs use a different strategy; they release a very fine thread of silk, which catches the breeze, hauling the spider away. Ballooning usually carries the spider for a small distance, but sometimes the spider may drift for many hundreds of miles. A few spiders are aquatic; the European water spider (*Argyroneta aquatica*) swims by

*A female* Argiope bruennicchi *in the hub of her web. Orb spiders spin radial webs, then wait for vibrations to reach their feet and alert them to the presence of struggling prey.*

beating its legs, while fishing spiders (genus *Dolomedes*) row across the water's surface at low speeds, using the front four legs. At higher speeds they break into a gallop, pushing against the surface tension of the water, and use all eight legs.

## Capturing prey

Fishing spiders often feed on fish and tadpoles, and some of the larger mygalomorphs occasionally eat larger vertebrates, such as birds and lizards. Most other spiders, however, feed exclusively on insects and small invertebrates, with a few crab spiders supplementing their diet with nectar. The prey of a spider may be equipped with chemical defenses, stings, or bites, and must therefore be subdued quickly and safely. To do this, most spiders use venom injected through the fangs (the exception are the orb-web building spiders of the family Uloboridae, which immobilize prey with silk before crushing it with the chelicerae). Spider venom is manufactured in glands in the cephalothorax. These are surrounded by rings of muscle that swiftly pump the venom out. The venom usually contains fast-acting neurotoxins, which block ion channels in the nervous system of the prey. The amount of venom injected can be controlled, with greater quantities being released to subdue especially large prey or a struggling victim.

Digestion of food starts outside the body, once the prey has been paralyzed. The spider injects enzymes through the fangs, which liquefy the internal organs of the prey. The resultant liquid is then sucked back up by the spider, and digestion is completed in the midgut.

Some spiders produce venom of staggering toxicity, which is far more powerful than that required simply for prey capture. This is usually for defensive reasons; one of the most venomous spiders known is the male Sydney funnel-web (*Atrax robustus*), which must wander far and wide searching for females. The male aggressively attacks if threatened, his venom nullifying any threats from potential predators. By contrast, the venom of the sedentary female, who is less vulnerable to predation, is only one-fifth as toxic.

## Stealth and silk

A number of spiders hunt using stealth. Crab spiders can change color to match their background, allowing them to mount an ambush—on, for example, brightly colored flowers that attract pollinating insect prey. Many others make silken webs and wait for insect prey to blunder in. Spider silk is elastic, extensible and very strong. Manufactured in glands in the abdomen, it is extruded from the spinnerets and has various uses in addition to prey capture. Silk is used to build retreats, or as a safety line should the spider have to make a quick escape. There are two main types of capture silk. Sticky silk has viscous globules along its length that trap prey through surface tension, while hackled band silk is coated with fine fibers, to which the hairs of an insect adhere. Orb-web spiders, such as garden spiders (genus *Araneus*), have a circular web, with a series of structural scaffold strands interspersed with sticky threads. The spider resides just above the hub, from which it can reach all parts of the web quickly. Many spiders use more silk to bind the struggling insect from range until it is safely wrapped and a paralyzing bite can be administered with minimal risk.

Many webs have zigzags of thickened silk across them. These are called stabilimenta. They are highly reflective to ultraviolet (UV) light, as are many flowers, and they may be used to lure in pollinating insects.

Webs are effective, but they are energetically costly to produce—even though many web-builders offset the energy costs by eating their web when the time has come to renew it. Some spiders minimize the energy outlay by capturing prey using fewer threads. Net-casting spiders (Deinopidae) hold a few strands of sticky silk between the front legs and swipe at passing insects, while the bolas spider (genus *Mastophora*) lures male moths with chemicals before lassoing them with a blob of viscous fluid at the end of a single line.

## Defense

Spiders have a range of enemies. The greatest predators of spiders are other spiders, often larger individuals of the same species. There are also a number of specialist spider-hunters, including the familiar cellar spider (*Pholcus phalangioides*). Another renowned arachnophage is the jumping spider *Portia*, which mimics the struggles of insect prey as well as the web plucking of potential mates in order to trick its victims into investigating. Parasitic wasps are also noted spider-hunters, although they do not feed on them; rather they paralyze their hapless victims, and lay an egg inside. The wasp larva feeds

*A female wolf spider carrying her young as she hunts. If they fall off, the spiderlings simply clamber back by a silken line.*

and develops inside the still-living spider's body. Some of the largest parasitic wasps can be found in the southwestern deserts of the United States.

Spiders use a range of anti-predator strategies. Many hide away in the daytime, feeding after dusk. Among the day-feeding species, some are camouflaged to blend in with bark, leaves, or even bird droppings. Perhaps the most remarkable defense strategy is that adopted by the ant-mimic spiders, family Clubionidae. Most animals give ants a wide berth, and the spiders exploit this by staying close to an ant colony. These spiders look very similar to ants; they have adopted the ants' characteristic running style and hold the front legs aloft to resemble antennae, so they walk on six legs rather than eight.

## Reproduction and growth

In order to breed, spiders often find each other by using pheromones. Sometimes silk is impregnated with these chemicals and incorporated into the web. Other species simply release pheromones directly into the air. Males are usually smaller than females, and run the risk of being viewed as prey rather than a potential mate. To counter this, many spiders have elaborate courtship rituals, involving the plucking of webs, drumming on the ground, and direct stroking with the palps. Male spiders from several families lightly wrap the female in a thin binding of silk, while the male nursery-web spider (*Pisaura mirabilis*) presents a freshly killed insect to the female. Toward the end of the breeding season, the female of certain species occasionally eats the male during or directly after copulation. This makes sense for the female, because the

*A beautiful specimen of* **Megaphobema** mesomelas *from Costa Rica. Such large, hairy species are commonly sold in pet stores as tarantulas, but they are in fact mygalomorphs.*

extra nutrients provide her with a better chance of producing eggs. If successful in charming a female, a male spider collects sperm on the modified end segment of his palps and delivers it to her epigyne, a receptor on the underside of her abdomen. After mating, some male spiders block the entrance to the female's ovary by breaking off the tip of the palp; this ensures that no further males will mate with her.

Eggs are often laid in a silken cocoon, and some spiders carry the eggs about with them in their chelicerae. Others attach the sac in a safe place and guard it from predators.

Many wolf spiders carry the spiderlings on their backs after they have hatched, while other females build a nursery nest. Some, such as the European mothercare spider (*Theridion sisyphium*), feed their young on regurgitated food. The young must shed their exoskeleton in order to grow, and spiderlings of larger araneids can molt up to 10 times before attaining maturity. Some female mygalomorphs continue to molt throughout their lives, which can last for over 20 years.

*For particular species see:*
- BIRD-EATING SPIDER • BLACK WIDOW
- CRAB SPIDER • DADDY LONGLEGS • FALSE SCORPION
- ORB SPIDER • PEDIPALPI • SCORPION
- SHEETWEB SPIDER • SOLIFUGID • TARANTULA
- TRAPDOOR SPIDER • WATER SPIDER • WOLF SPIDER

# SPINETAIL SWIFT

*Spinetails spend almost all of their lives on the wing. Pictured are several brown-backed needletails, among the fastest-flying of all swifts.*

SPINETAILS ARE A GROUP OF swifts with short tails, the feathers of which taper like those of a woodpecker so the central shaft protrudes beyond the vane. The spiny feathers are used as a prop when the swift is roosting on a vertical surface. There are about 25 species of spinetails, although some, for example those in the genus *Chaetura*, are not called spinetail. They are found in Africa, Asia and America, and one Asian species migrates to Australia and New Zealand. The best-known spinetail is the chimney swift, *Chaetura pelagica*, which breeds in North America. It is the common swift east of the Mississippi and Missouri Rivers and nests as far north as the Gulf of St. Lawrence. It winters in Central America and South America. The chimney swift is 4¾–5½ inches (12–14 cm) long, with a dark plumage but is whitish on the throat and upper breast. The spines on the tail feathers are not visible in flight and the tail is not fanned. The other North American spinetail is Vaux's swift, *C. vauxi*. Rather smaller than the chimney swift, it also has much paler underparts. It lives to the west of the Rocky Mountains, in the northwestern United States and southwestern Canada.

The other spinetails look similar to the North American species. Boehm's spinetail, *C. boehmi*, of Africa has a very short tail and broad wings, which give it a batlike appearance and account for its other common name of batlike swift. It is dark above, with a grayish brown throat and upper breast and a white lower breast and abdomen. The largest spinetail, and one of the largest of all swifts, is the brown-backed needletail, *C. gigantea*. This bird is 7 inches (17.5 cm) long and is dark with a whitish throat streaked with brown and a white abdomen. It ranges from India to Indonesia.

## Roosting in colonies

As with other swifts, spinetails do not land on the ground voluntarily. Their legs are small and weak and they are unable to spring into the air for takeoff. Consequently, they roost and nest on vertical surfaces, where they can take off merely by letting go and dropping away. The brown-backed needletail is one of the fastest flying swifts, and many of the spinetails fly at such great heights when feeding that they cannot be seen with the naked eye. However, the batlike spinetail, with its broad wings and short tail, is slower but can maneuver better than the others. Rather than hunting in great sweeps across the sky, the batlike spinetail hunts around the treetops in forests and woods. Its maneuverability is enhanced by long secondary flight feathers that perform the function of extra rudders.

Spinetailed swifts roost in colonies, sometimes numbering several hundreds, in chimneys, hollow trees and in the open. Regular roosts in chimneys and trees are used year after year. Each evening the swifts fly above the roost in a dense mass and then stream down into the roost.

Spinetail swifts eat insects caught in flight. These are mainly flies, bugs, mayflies and stoneflies. Some nestling chimney swifts were once found to have been fed on fleas, although it is not known where these came from.

## Time-saving nests

Courtship takes place on the wing, after which the pair of spinetails prospect for a nest site and mate there. With the exception of the large brown-backed needletail, which nests at the bottoms of tree hollows, spinetails build their nests on vertical surfaces in hollow trees, large chimneys and, in the case of Boehm's spinetail, occasionally in mine shafts.

The nest is made out of twigs that are broken off with the feet and carried in the bill, then cemented together with saliva in order to form a

# CHIMNEY SWIFT

| | |
|---|---|
| CLASS | **Aves** |
| ORDER | **Apodiformes** |
| FAMILY | **Apodidae** |
| GENUS AND SPECIES | ***Chaetura pelagica*** |

**WEIGHT**
**About ¾ oz. (21 g)**

**LENGTH**
**Head to tail: 4¾–5½ in. (12–14 cm)**

**DISTINCTIVE FEATURES**
**Small size; dark gray-brown body, paler on throat; torpedo-shaped form; pointed wings**

**DIET**
**Almost exclusively flying insects**

**BREEDING**
**Age at first breeding: 1 year; breeding season: eggs laid May–July; number of eggs: 4 to 5; incubation period: 19–21 days; fledging period: 30 days; breeding interval: 1 year**

**LIFE SPAN**
**Up to 14 years**

**HABITAT**
**Breeding season: wide variety of habitats; most associated with human habitations. Winter: forest, forest edge and scrub up to 9,900 ft. (3,000 m).**

**DISTRIBUTION**
**Breeding season: U.S. north to southernmost Canada east of Rocky Mountains. Winter: western South America.**

**STATUS**
**Common or abundant**

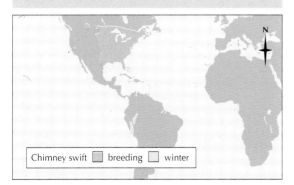

Chimney swift ▢ breeding ▢ winter

bowl-shaped structure. Fresh spinetail saliva has the consistency of glycerine and sets to form a tough solid in a few hours.

Egg-laying, in the chimney swift at least, starts when the nest is only half built. This appears to be a time-saving device, to ensure the nest is

completed before the chicks start to put on weight. The clutch of the chimney swift is two to seven eggs, three to seven in Vaux's swift, and they are incubated by both parents for 19–21 days. The chicks are brooded at night for another 12 days. They leave the nest at 14–19 days but do not take their first flight outside until they are 30 days old. They spend the intervening time clinging with strong claws to the wall of the hollow or chimney.

Very high levels of breeding success have been reported with chimney swifts. In one study there was an overall breeding success rate of 86 percent, with an average of slightly more than three and a half young per breeding attempt. First-year mortality is also low. The oldest ringed chimney swift was 14 years old.

## New nesting places

Before Europeans settled in North America the chimney swift nested and roosted in hollow trees. As industrialization spread, the bird began to use large chimneys and now it is very unusual to find a chimney swift nest in a tree. The new habit has allowed the chimney swift to extend its range into new regions where there are no trees.

*Altogether, five species of spinetails, including the chimney swift (above), have now acquired the habit of nesting on buildings.*

# SPINY ANTEATER

*A foraging echidna may use its snout to plow into a termite mound, or use its claws to rip into termite-riddled timber and expose the insects.*

THIS STRANGE-LOOKING CREATURE is, like the platypus (*Ornithorhynchus anatinus*), an egg-laying mammal of Australasia. It is also known as the echidna, native porcupine and even spiny porcupine. It resembles an oversized hedgehog, the back of its squat body sporting sharp spines over 2 inches (5 cm) long. Like a hedgehog, but even more so, it is powerful.

There are two genera: *Tachyglossus*, the short-beaked echidna, and *Zaglossus*, the long-beaked spiny echidna. Each contains a single species. The latter is the larger, weighing up to 22 pounds (10 kg). Its very long snout houses a similarly long tongue. The legs are much longer than in the *Tachyglossus* species.

In the short-beaked echidna the head and body are up to 18 inches (45 cm) long, and the tail is short, stubby and naked and only about 3 inches (7.5 cm) long. It may be over 10 pounds (4 kg) in weight, and there is a record of one topping 14 pounds (5.7 kg). The spines are usually yellow at the base and black at the tip, but they are sometimes entirely yellow. They are interspersed with coarse brown hair, which extends over the underbelly. The extent to which hair is visible through the spines depends partly on latitude and climate: echidnas in warmer northern regions are more spiny, while animals in cooler southern parts, such as Tasmania, can be quite hairy. This variation is not considered sufficient to support the naming of subspecies.

There is no neck, and the small external ears are not usually visible. The small eyes are at the base of the long, tapering snout, which has a small, toothless mouth. The long, saliva-coated tongue can be shot out as much as 6–7 inches (15–17.5 cm). The legs are stocky, and each of the immensely strong forefeet has five long, curved claws for digging. The second of the five toes on a hind foot is long and is used for grooming. All males and some females have short ankle spurs.

## Powerful digger

Spiny anteaters live in a wide variety of habitats ranging from hot dry deserts and humid rain forests to rocky ridges and valleys, up to an altitude of 5,000–6,000 feet (1,520–1,830 m) in the Australian Alps, where the air temperature during the three coldest months of the year seldom rises above freezing. The New Guinea species of *Zaglossus* inhabit humid forests at altitudes of 3,770–9,400 feet (1,150–2,870 m).

# SPINY ANTEATERS

| | |
|---|---|
| CLASS | **Mammalia** |
| ORDER | **Monotremata** |
| FAMILY | **Tachyglossidae** |

**GENUS AND SPECIES** **Short-beaked spiny anteater,** *Tachyglossus aculeatus;* **long-beaked spiny anteater,** *Zaglossus bruijni*

ALTERNATIVE NAMES
**Short-beaked echidna; long-beaked echidna; short-nosed echidna; long-nosed echidna; native porcupine; spiny porcupine**

WEIGHT
**5½–22 lb. (2.5–10 kg)**

LENGTH
**Head and body: 14–31 in. (35–78 cm)**

DISTINCTIVE FEATURES
**Long snout; thickset legs with long, curved claws; covered in spikes and fur**

DIET
**Ants, termites, worms and other invertebrates**

BREEDING
**Age at first breeding: 1 year; breeding season: peaks July–August; number of young: 1 to 6; period in pouch: 6–8 weeks; breeding interval: 2 years**

LIFE SPAN
**Up to 20 years**

HABITAT
**Plains, rocky areas, alpine meadows, forests**

DISTRIBUTION
**Short-nosed echidna: throughout Australia and Tasmania, also in coastal areas of New Guinea; long-nosed echidna: confined to parts of New Guinea highlands**

STATUS
**Short-nosed echidna: common; long-nosed echidna: endangered**

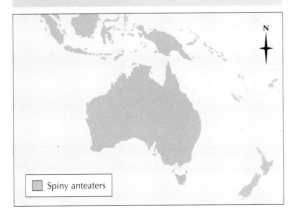

Spiny anteaters

The echidna is generally solitary. It can run swiftly and climb well. It walks with the legs fully extended, so that the stomach is relatively high off the ground and the hind toes are directed outward and backward, giving it a somewhat awkward appearance. It shelters in burrows or crevices among rocks, and when disturbed it digs into the soil at great speed, descending vertically so that the spines are uppermost. If the soil is too hard for digging, the echidna rolls up, presenting an armory of spines to a potential aggressor.

Naturalists once believed the echidna to be nocturnal, but it is now known to be active at all hours of the day, especially on cool afternoons. It spends much of its time, however, rolled up, presumably asleep, when not actively foraging. The long-beaked echidna has the lowest body temperature of any mammal and a slow metabolism to match; it has no sweat glands, and to raise its body temperature above 95° F (34° C) can be fatal. Accordingly, the echidna avoids extreme heat by hiding in hollow logs or rock crevices, and may even take a swim to cool off.

## Licking up its food

A spiny anteater eats both ants and termites, though it prefers the latter, which are meatier and less fierce in defense. Moreover, it forages in the warmer, southern aspect of a termite mound, where the queen is most likely to be located. The echidna licks up the prey with a long tongue coated in saliva as thick as molasses. It may simply lay the tongue over a mound, or thread it down a gallery and around U-bends with ease.

The echidna's jaws are toothless. Instead of chewing, it scrapes insects from the tongue onto spines on the palate, and crushes them between

*Using its snout and immensely powerful limbs, a spiny anteater can burrow beneath the surface in a matter of seconds.*

these spines and another set on the base of the tongue as it is thrust out again. Eating in this manner, an adult echidna can devour 6 ounces (200 g) of ants in 10 minutes.

The long-beaked echidna likes earthworms and has a specially modified tongue, tipped with spikes, for grasping them. Aligning them with its snout, the echidna then sucks them down. The spiny anteater is said to be able to forgo food for up to a month. Like a hedgehog, a tame echidna will eat bread, milk and minced meat.

## Egg-laying mammal

Long-beaked echidnas mate from early July to late September. The courtship involves several males as they follow a female around, tracking her by scent. When she is ready to mate, her suitors gather around, and each tries to push the others out of the vicinity. The victor will mate with her. Three to four weeks later, the female lays one egg, occasionally two, directly from the cloaca (waste and genital opening) into a pouch that will have opened on her belly at the start of the breeding season. The long-beaked species may lay up to six eggs in the breeding season.

Each egg is grape-sized and spherical, with a soft, parchmentlike covering. It hatches in 10–11 days. The baby, known as a puggle, breaks out by hacking at the shell with its egg tooth. It is suckled within the pouch on milk that is secreted from slitlike ducts in the mother's abdomen; pinkish and iron-rich at first, the milk has a formula unlike that of any placental or marsupial mammal. The puggle, which grows rapidly, remains in the pouch until its spines are sufficiently developed that the mother must eject it. The pouch later disappears.

The mother then deposits her offspring in a burrow and visits it every 1½–2 days to feed it until it is weaned about three months later, when it weighs about 2 pounds (800 g). An echidna becomes sexually mature at the end of one year and, in captivity at least, may live up to 50 years.

## Mixed fortunes

The long-beaked echidna has few natural enemies. Aborigines eat it, and the flesh is said to have a characteristic smell of crushed ants. Fortunately, it is never a pest and it does not have any economic values for humans to exploit. As such, the species is in no danger.

The same is sadly not true of the long-beaked echidna. Its highland forest habitat in New Guinea is under increasing pressure from human settlers, who, in addition to stripping the land, also hunt the echidna for its rich, oily flesh. Unless conservation measures are established, this species may soon face extinction.

*Echidnas have a sixth sense, an ability to detect electrical signals through sensors in the snout. This may help them detect the activity of invertebrates deep in the ground.*

# SPINY EEL

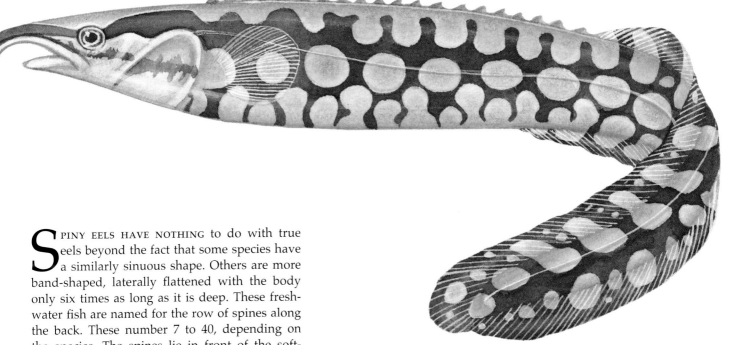

SPINY EELS HAVE NOTHING to do with true eels beyond the fact that some species have a similarly sinuous shape. Others are more band-shaped, laterally flattened with the body only six times as long as it is deep. These freshwater fish are named for the row of spines along the back. These number 7 to 40, depending on the species. The spines lie in front of the soft-rayed dorsal fin and can be raised or lowered at will. Some spiny eels, when picked up, wriggle backward with the spines erect, which can be extremely painful to the handler.

Most of the 71 species are less than 15 inches (37.5 cm) long, the longest being 3 feet (90 cm). In color they are mainly some shade of brown on the back and yellow to pale brown on the belly. This muted background coloration is enlivened by attractive patterns of spots, irregular stripes or patterns along the flanks. The head tapers sharply to the snout, which curves slightly down. The double nostrils, found in so many fish, are widely separated. The front nostril opening is tubular and hangs down near the tip of the snout, while the rear opening lies just in front of each eye. The mouth and gill openings are small. The dorsal and anal fins start toward the middle of the body and are usually continuous around the tail tip. The pectoral fins are small, and the pelvic fins are absent.

Spiny eels live in the fresh and brackish waters of tropical and southern Africa; their range extends north to southern Asia and southern China and eastward to the Malay Archipelago.

## Mud dwellers

Spiny eels live in both still and running waters with a muddy or sandy bottom and with plenty of aquatic vegetation. In parts of southern Asia and Africa they occur in rice paddy fields, swamps, shallow coastal plains and estuaries. They swim with side-to-side, serpentine movements of the body, assisted by a wavelike action of the fins. During the day they lie hidden among water plants or buried in the mud or sand. Some species dig themselves in until only the eyes and nostrils are exposed at the surface. To do this they rock the body from side to side, at the same time thrusting forward with the snout. Many breathe air, rising to the surface to gulp it, and these fish can survive indefinitely in foul waters deficient in oxygen.

## Hog-nosed feeding

In the evening a spiny eel leaves its daytime hideout to feed. The long, mobile snout, which is strengthened with a rod of cartilage, is highly sensitive. With it, and helped by the tubular nostrils, the spiny eel feels around for prey and then sucks it in with a jerky movement. It also forages in the mud, rooting into it like a hog. Its diet is composed mainly of worms, insect larvae and small crustaceans.

## Hide-and-seek fry

At the onset of the breeding season, male spiny eels chase the females, which are the more robust of the two sexes. Males use their snouts to nudge

*A row of sharp, erectile spines in front of the soft, hind part of the dorsal fin gives these intriguing fish their common name.*

the females in the vent region. The eggs appear to be scattered over the bottom. Little is known about development, although observations of one species revealed that the eggs hatched in 3 days. The fry, ignored completely by the adults, immediately hid themselves among the thickest tangles of plants or buried themselves in the soft surface of the mud. Young spiny eels lack the long snout, which does not begin to show until they are at least 1 month old.

## Slippery customers

Little is known about the natural predators of spiny eels, though the fry are assumed to be on the menu of most larger aquatic life. The adults are difficult to spot, being the same color as the mud in which they hide. Moreover, many have a disruptive pattern on their bodies that tends to break up their outline, rendering them inconspicuous as they swim about in broad daylight among water plants. From the experience of the few who have tried to collect them, the eels reportedly dive into the mud as soon as they are disturbed. Nevertheless, some species are caught and eaten. These include the blind spiny eel (*Mastacembelus brichardi*) of the Congo River in equatorial Africa, the one-striped spiny eel (*Macrognathus aral*) of southern Asia, and to a lesser extent the barred spiny eel (*M. pancalus*) also of southern Asian lowlands. In certain places spiny eels are avoided because of their resemblance to snakes. They are harmless, however, having no venom and only tiny teeth in their very small mouths.

## Mystery snout

The lesser spiny eel, *M. aculeatus*, looks more like a caricature than a real fish. It has been called the elephant trunk fish. It has a long, highly mobile snout that is partially prehensile. On the underside of the snout are 20 to 26 pairs of toothed plates carried on a forward extension of the upper jaw. How it uses this specialized snout is not clear, but its appearance suggests that the fish uses it in the same way as an elephant uses its trunk—to transfer food to its mouth.

## Favorite among aquarists

Spiny eels make an unusual feature in the aquarium, and several species are collected for live trade, including the tiretrack spiny eel (*Mastacembelus armatus*), so named because of its zig-zag markings; the fire eel (*M. erythrotaenia*); and the peacock eel (*M. circumcinchus*). The barred spiny eel has been bred in captivity. Spiny eels are said to thrive in water temperatures of 78–83° F (26–28° C), and can reportedly be trained to accept food from the hand. They are, however, pH-sensitive and dislike abrupt water changes.

---

## SPINY EELS

CLASS **Osteichthyes**

ORDER **Synbranchiformes**

FAMILY **Mastacembelidae**

GENUS AND SPECIES **Aethiomastacembelus (20 species), Caecomastacembelus (29 species), Macrognathus (12 species), Mastacembelus (9 species, including M. armatus, detailed below) and Sinobdella (1 species)**

ALTERNATIVE NAME
**M. armatus: tiretrack spiny eel**

WEIGHT
**Up to 17⅔ oz. (500 g)**

LENGTH
**Up to 30 in. (75 cm)**

DISTINCTIVE FEATURES
**Eel-like body with long head; brown above, yellowish below; fleshy, mobile snout; small mouth and gill openings; small scales over body; dorsal fin is a row of isolated, erectile spines followed by long, soft-rayed fin; irregular, dark, zig-zag band from eye to root of tail, branching off to dorsal and anal fins; young often have cloudy blotches**

DIET
**Insect larvae, worms and some water plants**

BREEDING
**Males court females, which scatter eggs on bed; fry develop among weeds or mud**

LIFE SPAN
**Not known**

HABITAT
**Rivers with sand, pebble or boulder beds; also still waters, including coastal marshes**

DISTRIBUTION
**Southern and Southeast Asia, from Pakistan, Bangladesh and Myanmar (Burma) to China**

STATUS
**Locally common**

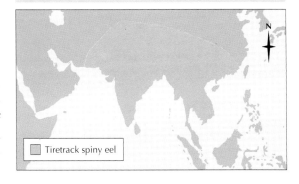

Tiretrack spiny eel

# SPINY MOUSE

THE SPINY MOUSE OF Africa and Asia is another example of a mammal with a spiny coat whose sharp prickles have evolved from soft hairs, presumably for self-protection. There are 14 species ranging from 2¾ inches (7 cm) to nearly 7 inches (17.5 cm) in head and body length, with a slightly smaller or equal-length tail. The average weight of an adult of the larger species is 2–3 ounces (56–85 g). The color on the upperparts varies from pale yellow, reddish brown and reddish to dark gray, whereas the underparts are white. Some species produce melanistic (all-black) individuals.

The back and tail are covered with coarse, rigid grooved spines. The tail, which is scaly and near-naked, is brittle and is easily broken. It may be whitish above and below, entirely dark or bicolored. The snout is pointed, the ears are large and erect and, as in most mice, the eyes are prominent and bright.

A typical member of the genus *Acomys* is the Cairo spiny mouse (*A. cahirinus*), named from specimens found in Cairo early on in the 19th century. The remaining species range across much of Africa and from India through southwestern Asia. These include the golden spiny mouse (*Acomys dimidiatus*) from northeastern Africa and southwestern Asia, which is distinguished by the black soles on its feet, and the Cape spiny mouse (*A. capensis*) of South Africa, which is less prickly than the typical species.

## Thrive in arid habitats

Spiny mice usually live in rocky, arid country, but in parts of their range they are found in woodland undergrowth and desert areas. In deserts they often use gerbil burrows for shelter, and when food is scarce they may compete with gerbils for food. They usually come out to feed in the early morning and late afternoon but individuals have been seen about at all hours.

In the wild spiny mice eat mainly seeds and other plant foods, including dates. Some of them live in or near human habitations, however, and these, like the house mice, feed on nearly anything they can find. They have been known to chew such materials as fiber mats, and there have even been reports of Cairo spiny mice nibbling on mummified corpses in tombs on the banks of the Nile at Asyut, Egypt. (In that country, spiny mice have also been implicated in transmitting typhus.) In desert areas spiny mice have been observed feeding on snails.

## Active young

Though breeding can occur at almost any time of year, the young of most spiny mice are born between February and September.

The Cairo spiny mouse shows unusual breeding behavior. At about 5 weeks long, its gestation period is substantial for such a small animal. Just before giving birth the female shows an exaggerated maternal urge and often steals a neighbor's baby, holding it in her paws to groom it, and she may even suckle it. In the absence of any young ones around, she may even attempt to groom an adult mouse, trying to hold it in her paws as if it were a baby. The female does not retire into a quiet corner when about to give birth and actually gives birth standing up, the young generally being delivered backward as among large hoofed animals. This is unusual in rodents.

*Young spiny mice stick close to their mother for warmth, food and protection from predators. They are weaned at 2 weeks.*

*A female spiny mouse may act as wetnurse to the young of another mother, simply gathering the extra litter under her body and suckling them alongside her own.*

## SPINY MICE

| CLASS | Mammalia |
|---|---|
| ORDER | Rodentia |
| FAMILY | Muridae |
| GENUS | *Acomys* |
| SPECIES | 14 species |

**LENGTH**
Head and body: 2¾–7 in. (7–17.5 cm); tail: 1½–5⅕ in. (4–13 cm)

**DISTINCTIVE FEATURES**
Coat brown above, white below; short, stiff spines on back and tail; tail otherwise naked and scaly; large ears, black eyes

**DIET**
Grasses and seeds; some insects and carrion

**BREEDING**
Age at first breeding: 2–3 months; breeding season: variable; gestation period: 35–42 days; number of young: 1 to 5; breeding interval: produces up to 12 litters in a row

**LIFE SPAN**
Up to 5 years

**HABITAT**
All arid habitats

**DISTRIBUTION**
Africa, Sicily, Crete, Cyprus, Middle East and southwestern Asia

**STATUS**
Most species very common; 1 species in Turkey considered critically endangered; 2 species on Crete listed as vulnerable

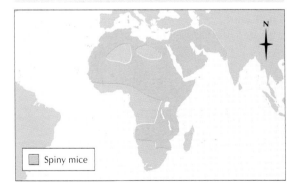

Spiny mice are unique among the family Muridae in that the young of most species are born fully furred and with their eyes open, showing none of the helplessness of most newly born rodents. The exception is *A. dimidiatus*, whose young are born blind, their eyes opening on the third day. A curious habit of licking their mother's saliva has been observed in the young; scientists do not know the reason.

Weaning occurs about 2 weeks after birth, when the young weigh about ½ ounce (14 g). Males are sexually mature at about 7 weeks old. Their growth is almost complete at the end of about 150 days, although weight tends to increase slightly up to the third year. A curious phenomenon, for which there is no ready explanation, is that shortly before the death of a spiny mouse the animal becomes slightly emaciated and there are dramatic changes in its weight.

## Other spiny mice

True spiny mice, genus *Acomys*, are found only in Africa and Asia. There are, however, similarly named species in other families. In tropical America, from Guyana and Brazil to Panama, there are four species of so-called spiny mice, but these belong to Cricetidae, the vole family. Another species, in Ecuador, is known from only six specimens collected in 1924 and has not been seen since. Found from Mexico to Texas are 21 species of spiny pocket mice of the family Heteromyidae. Apart from their stiff coats, they differ little from ordinary voles and pocket mice, and their coats are not all equally spiny, some being covered in little more than stout bristles.

# SPONGE

THE SPONGE SEEN IN the bathroom is often, in fact, the fibrous skeleton of an animal. In life the gaps in the skeleton are filled with a yellowish flesh, the whole organism being covered with a dark purple skin. A sponge has no sense organs and few specialized organs apart from chambers containing collared cells. At best, there are a few scattered muscle cells and very simple nerve cells.

Although there are nearly recognized 3,000 species of sponges, the most familiar is the group of half a dozen species of horny sponges in the genus *Spongia*. On the market, bath sponges are given common names, such as fine turkey, brown turkey, honeycomb for Mediterranean sponges, and wool, velvet, reef, yellow and grass for Bahamas sponges. These names express mainly the varying textures of the fibers. There are many other names, but the fine turkey and the honeycomb are those most often seen for sale. They are found only in warm seas, to depths of 600 feet (180 m). They are most numerous in the Mediterranean, particularly the eastern part, and off the Bahamas and Florida. Elsewhere, in tropical and subtropical waters, sponges are found of similar type but less durable or pleasing texture, although in places, as in Southeast Asia, there may be limited fisheries supplying a local market.

*Crowding the surface of a Red Sea sponge are many small gemmules, or asexually produced offspring. Also visible are two of the vents from which waste water is expelled after feeding.*

## Sitting pretty

A sponge normally draws all it needs from the sea without departing from the spot on which the larva settled. The beating of the protoplasmic whips, or flagella, of its collared cells, which are grouped in rounded chambers in the network of canals running through the body, draws in currents of water through many minute pores in the skin. Having passed through the chambers, the water is driven back toward the surface and expelled with moderate force through craterlike vents that are usually larger and less numerous than the inlet pores. In its course through the sponge, the water yields food particles and oxygen, and it ferries away waste products of digestion and respiration.

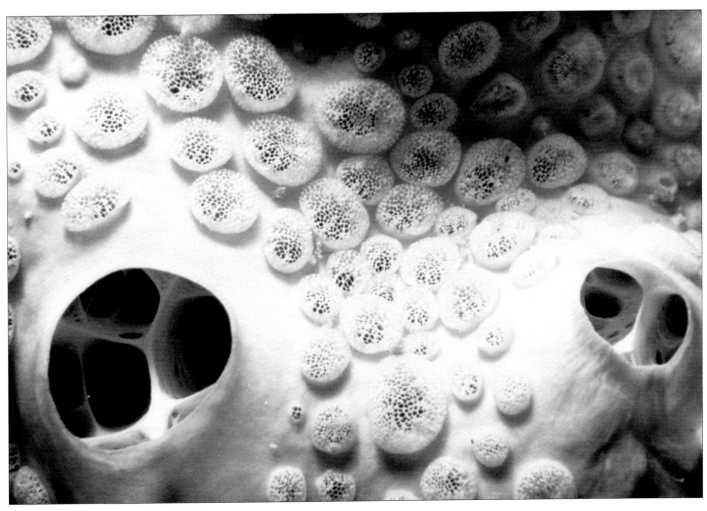

## BATH SPONGE

| | |
|---|---|
| PHYLUM | **Porifera** |
| CLASS | **Demospongiae** |
| ORDER | **Dictyoceratida** |
| FAMILY | **Spongidae** |
| GENUS AND SPECIES | ***Spongia officinalis*** |

SIZE
**Up to 20 in. (50 cm) across; normally about 12 in. (30 cm)**

DISTINCTIVE FEATURES
**Pale brown to yellow color; many holes of various sizes on surface; spongy texture**

DIET
**Filter feeder on tiny organic particles**

BREEDING
**Each sponge produces both eggs and sperm; sperms released into water and inhaled by another sponge for internal fertilization to occur; tiny larvae, released a few days later, spends hours or days swimming with plankton before settling and developing into new sponge**

LIFE SPAN
**Probably up to 30 years or more**

HABITAT
**Rock surfaces to depths of 150–180 ft. (45–55 m) in warm waters**

DISTRIBUTION
**Coastal waters of Pacific, Indian and Atlantic Oceans; also in Mediterranean**

STATUS
**Locally common; also cultivated**

*A mature sponge in the Caribbean. After several years sponges may become colonized by other marine animals, such as anemones. Some crabs also pick up small sponges to use as shell camouflage.*

Sponges are almost completely sedentary once the free-swimming larva has settled on a solid substrate—usually a steady boulder. There is, however, evidence of limited movement, specially in young sponges, their speed being about 1 inch (2.5 cm) in 2 weeks. The movement may be a response to adverse conditions.

### Particulate feeders

Sponges feed on bacteria and minute particles from the breakdown of plant and animal bodies. Sponges are therefore particulate feeders and scavengers. Because sponges have no separate digestive system, food particles are taken into the collared cells, which digest them and reject any inedible scraps. The food is then passed into the body by migrating, amoeba-like cells.

### Ciliated larva

There is no such thing as a male or a female sponge: each sponge produces eggs and sperm. At various points in the body of the bath sponge, *Spongia officinalis*, one of the body cells is fed by neighboring cells so that it grows noticeably large. Of the many thousands that undergo this process, some are destined to become egg cells.

The other cells subdivide repeatedly until masses of tiny cells are formed. These are the sperms. When ripe, each bursts from its capsule into the water canals and escapes by the vents into the sea. The sperms swim around until, drawing near another sponge, they are sucked in by the water current entering through its pores. Inside, they travel through the canals until they meet an egg, which one of them fertilizes.

The fertilized egg divides repeatedly to form an oval mass of cells: the embryo. Some of these put out flagella, and as they beat, they cause the embryo to rotate. This breaks its capsule, and the embryo, now a free-swimming larva, swims out through one of the vents. For the next 24 hours it swims in a spiral motion with its flagella. Then the flagella weaken, and the larva sinks to the seabed. There, it develops into a small platelet of tissue the size of a pinhead. This is a new sponge.

Sponges also reproduce by asexual means, in which gemmules, budlike aggregates of cells, form on the surface of the parent sponge.

## From seabed to bath

From the moment that the larva changes into the pinhead-sized sponge to the time this has grown large enough to be put on the market, seven years must elapse. By then it has developed into a sponge the size of two clenched fists. If left to its own devices, a sponge may live much longer and grow considerably larger. Bath sponges 20 inches (50 cm) in diameter have been brought up. These were probably 20 or more years old.

If a living bath sponge is cut into two, each piece heals the cut and grows into a new sponge. The same powers of regeneration will take place in a sponge cut into 4, 6, 12 or even more pieces: it produces 4, 6, 12 or more new sponges in due course. More than a century ago Oscar Schmidt, an Austrian zoologist, proposed that sponges be grown from cuttings, like plants. Early in the 20th century the idea was adopted by the British Colonial Office. Experiments in growing sponges from cuttings were carried out in the Bahamas and in the Gulf of Mexico. Each sponge was fastened to a concrete disc, and all were laid out in rows on the seabed.

Misfortune dogged the experiments. One year, when rows of cuttings had been laid down, a hurricane swept them all away. The farmers persevered, and in 1938 some 600,000 tons (545,000 tonnes) of sponges were harvested from the Gulf of Mexico alone. Though sponges have few predators except nudibranches (sea slugs), they can suffer from a fungal disease. This struck in 1938, causing 90 percent mortality among sponges in the Bahamas and Florida.

*A diver investigates a large barrel sponge, genus* Petrosia, *off Papua New Guinea. Sponges commonly settle on submarine cliffs such as these.*

# SPOONBILL

THE SPOONBILLS, WATERBIRDS related to the ibises, are remarkable for the peculiar shape of their bills. The six species of spoonbills are up to 3 feet (90 cm) long from the tip of the bill to the end of the tail and they have long legs. The bill is long, flattened horizontally and broadened to a spatulate or flattened spoon shape at the end. Whereas the ibises belong to many different genera, all of the spoonbills are members of the genus *Platalea*.

## Strange bills, graceful plumage

Spoonbills have almost wholly white plumage except for the roseate spoonbill, *Platalea ajaja* (formerly called *Ajaja ajaja*), of North, Central and South America. The roseate spoonbill's adult plumage is strongly tinged with pink, deepening to red on the shoulders. Its legs are pink and its bill is yellow. The roseate spoonbill ranges from the southern United States through much of

South America. The Eurasian spoonbill, *P. leucorodia*, has a black bill tipped with yellow, black legs, an orange-yellow patch on the throat and a yellowish patch at the base of the neck in summer, when it also has a drooping crest at the back of the head. In winter it loses its crest and the yellow areas. The Eurasian spoonbill breeds in parts of southern and eastern Europe and across Central Asia to China, as well as in the Netherlands, the coastal regions of northern and East Africa and the valley of the Nile River.

The lesser spoonbill (*P. minor*), of China and Japan, is very similar to the Eurasian spoonbill, as are the royal or black-billed spoonbill (*P. regia*) and the yellow-billed or yellow-legged spoonbill (*P. flavipes*), both of Australia. Some ornithologists used to be inclined to the view that these four Old World spoonbills are just geographic races, or subspecies, of the same species. However, most authorities now accord them full

*A roseate spoonbill bathing to keep its plumage in top condition. Found from Florida and Texas to South America, this is the most colorful species of spoonbill.*

# SPOONBILL

| | |
|---|---|
| CLASS | **Aves** |
| ORDER | **Ciconiiformes** |
| FAMILY | **Threskiornithidae** |
| GENUS AND SPECIES | ***Platalea leucorodia*** |

**ALTERNATIVE NAME**
**Eurasian spoonbill**

**WEIGHT**
**2½–3⅓ lb. (1.1–1.5 kg)**

**LENGTH**
**Head to tail: 2⅔–3 ft. (80–90 cm);
wingspan: 3¾–4¼ ft. (1.15–1.3 m)**

**DISTINCTIVE FEATURES**
**Long, flattened bill with very broad spatula
or spoon shape at end; heronlike body;
long black legs. Breeding adult: pure white
plumage except for golden yellow throat
and breast band; drooping white crest.
Nonbreeding adult: no crest or yellow areas.**

**DIET**
**Aquatic insects and their larvae, mollusks,
crustaceans, worms, leeches, small fish,
frogs and tadpoles; some plant material**

**BREEDING**
**Age at first breeding: 3–4 years; breeding
season: eggs laid April–May (Europe);
number of eggs: usually 3 or 4; incubation
period: 24–25 days; fledging period: 45–50
days; breeding interval: 1 year**

**LIFE SPAN**
**Up to 30 years**

**HABITAT**
**Shallow marshes, estuaries, lakes and pools**

**DISTRIBUTION**
**Eastern Europe east through Central Asia to
China, south to northern Africa, East Africa
and India; also in Spain and Netherlands**

**STATUS**
**Uncommon**

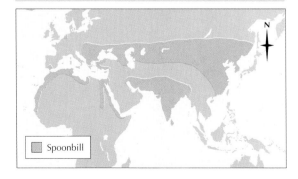

Spoonbill

species status. The African spoonbill, *P. alba*, has mainly white plumage but differs from all the rest in having bare red skin on the face, some red on the bill and bright pink legs. It lives throughout Africa south of the Sahara Desert and also in Madagascar.

## Sweeping for food

All spoonbills are alike in their habits except that the roseate spoonbill has less tendency to nest on the ground. They live together in loose flocks, often in the company of ibises, herons and egrets, all relatives of the spoonbills. They are found mainly near shallow fresh water or in marshes. Spoonbills swim only occasionally but perch in trees, often high up. Their flight is slow with regular wingbeats, and they sometimes glide and soar. A party of spoonbills flies in a single file at fairly regular intervals, with their long necks stretched forward.

During the day spoonbills rest and sleep by the edge of the water, sometimes on one leg. They may feed a little by day but they usually feed at dusk. A spoonbill holds its bill vertically downward and sweeps the spatulate end from side to side underwater, keeping the bill slightly open to grasp anything edible. At times the bird rushes here and there, backward and forward as if frantic, probably chasing an elusive quarry. Although a small amount of seeds and fibrous plant matter have usually been found in the stomachs of dead spoonbills, their food is mainly animal. It includes the usual insect larvae and aquatic insects as well as leeches, worms, water snails, frog spawn, tadpoles and the small fish that haunt the shallows.

*Spoonbills are often
seen in the company
of ibises, herons and
egrets. In this photo a
few Eurasian spoonbills
have formed a flock
with several sacred
ibises,* Threskiornis
aethopithecus, *which
can easily be identified
by their black heads.*

## Silent nesting

Spoonbills are generally silent birds, although they are said to give a low grunt when disturbed at their nesting places, and the nestlings are said to squeak and wheeze. When excited, as at breeding time, they use their bills as clappers, at the same time raising the feathers on their crest. Courtship seems also to be accompanied by the kind of dancing seen in storks, but this has only rarely been observed.

Spoonbills usually nest in reed beds or in low bushes, on islands in the marshy lagoons, but the roseate spoonbill more often nests in tall trees. The birds nest in colonies of a few to 200 pairs, the nests often being built only a few feet apart. Both of the pair build the nest, which is made of reeds or sticks piled to a height of 1½ feet (45 cm). Usually three or four eggs are laid. They are a dull white with sparse brown markings. The eggs are incubated for 24–25 days by both parents. The downy young remain in the nest for 4–5 weeks and return to it regularly for another 2–3 weeks by which time they are able to fly. At first each nestling has a common bill, but the spoon shape soon emerges. They are fed by the parents regurgitating food into their throats and then opening their bills wide and allowing the chicks to thrust their massive bills in to take the partly digested food.

## Damage from drainage

The main enemies of spoonbills are their own bills and the way in which they feed. Both are specialized to give the maximum return from feeding in the shallow waters of marshes and in shallow lagoons and lakes. These habitats are the first to go when land is drained to reclaim it for agriculture. For example, spoonbills used to breed in parts of England until the 17th century. Then the draining of marshes robbed them of the habitat necessary for their peculiar bills and ways of feeding.

The peculiar habitat requirements of spoonbills account for their highly localized and scattered distribution. In Europe, for example, Eurasian spoonbills are plentiful in the Netherlands in the west and in the Danube Basin in the east; they occur in very few places in between. Spoonbills must have extensive shallows of fairly even depth, with bottoms of mud, clay or fine sand, and preferably with gentle tidal changes or slow currents.

*Unlike the five other spoonbills, the African spoonbill has bright red legs and red skin on the face. It is a nomadic species, wandering in search of suitable feeding and breeding grounds.*

# SPRINGBOK

ALSO KNOWN AS springbuck or springbock, the springbok is a lively antelope and is the national symbol of South Africa. It lives in the Kalahari Desert of southwestern Africa, extending eastward as far as Botswana and Zimbabwe. The springbok extends also into the western side of the Republic of South Africa and into Free State, but is no longer as numerous as it once was in these areas.

The springbok's name is derived from its characteristic high-backed spring, or pronk. In this movement, the animal leaps 10–12 feet (3–3.6 m) into the air with the body curved and the legs stiff and close together, pointing downward; the head is held low.

One purpose of the pronk may be to alert nearby springbok to an intruder. However, the movement is mostly characteristic of young springbok and is probably a gesture that enables juveniles to orient themselves. By slowing down the speed of the bounce and maximizing its height, young springbok can take bearings on their surroundings and check on the position of their neighbors and any predators. Mature animals pronk less often, probably because they are more aware of their environment.

## Gazellelike in form

Although it resembles a gazelle in form, the springbok is not related to the gazelle family. It is larger than most gazelles, standing up to 36 inches (90 cm) high and weighing up to 152 pounds (69 kg). Typically reddish fawn in color, it is white underneath with a black stripe along the flanks separating the fawn from the white. The springbok has a mainly whitish face with only a narrow black stripe through the eye to the nose, partly obliterating the typical gazellelike face markings. The horns are ringed and slightly divergent with strongly inwardly curving tips, and they are smaller in females.

From the middle of the back to the rump is a large, pocketlike gland. When the springbok becomes excited, this turns inside out, revealing a broad display of erect, pure white hairs that act as a warning signal to other springbok. The first animal alerted shows these hairs, and the signal is taken up and passed on by others. The sight of the erect white hairs also tells a predator that it has been spotted. If it persists in pursuing the springbok, the predator may well be outpaced, as a springbok is capable of speeds of up to 55 miles per hour (88 km/h).

*In common with gazelles, the springbok has a striped face and bands on its thighs and flanks. The horns and body form are also gazellelike, although it is not actually a gazelle.*

*During the rut (mating season), which can happen at any time in the year, sexually mature male springbok become fiercely aggressive toward one another.*

It is likely that there are two primary reasons for the close resemblance between springbok and gazelles. Springbok share a common ancient ancestor with gazelles, which was very similar in form and size to its present-day descendants. Moreover, both have evolved very similar ecological patterns of behavioral patterns to enable them to survive in extremely dry habitats.

## Range widely

Springbok live in large herds, the social structure of which is not well known. The herd is probably centered on a few territorial males, each with its mixture of females and their young, as is the case in many gazelles. The males mark their territories with urine and feces and repel any intruding males. However, if the females opt to move to the territory of another male, in which food and water is more plentiful, the male can do little to prevent them. Younger males form bachelor herds that are composed of as many as 50 animals.

The dry habitat and patchy distribution of their food means that springbok must range over wide areas to find sufficient nourishment. They feed mainly on grasses, although they browse on bushes in the winter. At the arrival of rain, the small groups of springbok assemble into large herds and graze extensively on the newly grown grass and herbage. Springbok feed mainly at dawn and dusk.

When a predator, such as a leopard, lion, cheetah or wild dog, has been located and all nearby animals are alerted to its presence, the whole herd begins to run. With their bodies fully extended, they run low over the ground, in the manner of blesbok (discussed elsewhere).

## SPRINGBOK

| | |
|---|---|
| CLASS | **Mammalia** |
| ORDER | **Artiodactyla** |
| FAMILY | **Bovidae** |
| GENUS AND SPECIES | ***Antidorcas marsupialis*** |

ALTERNATIVE NAMES
**Springbuck; springbock**

WEIGHT
**44–152 lb. (20–69 kg)**

LENGTH
**Head and body: 4–5 ft. (1.2–1.5 m); shoulder height: 26¾–36 in. (68–90 cm); tail: 6–11 in. (15–28 cm)**

DISTINCTIVE FEATURES
**Gazellelike form, but only distantly related to gazelles; light brown back; white belly, head and some parts of lower flanks; black stripe across eye, along flank and next to white tail; inwardly curving horns**

DIET
**Mainly grasses; also leaves**

BREEDING
**Age at first breeding: 8 months, but male must wait until it wins a territory before breeding; breeding season: all year; number of young: usually 1; gestation period: 168 days; breeding interval: about 1 year**

LIFE SPAN
**Up to 20 years in captivity**

HABITAT
**Dry, open plains**

DISTRIBUTION
**Southern Africa, from northern Angola south to South Africa and east to Zimbabwe**

STATUS
**Conservation dependent; bred in captivity as livestock**

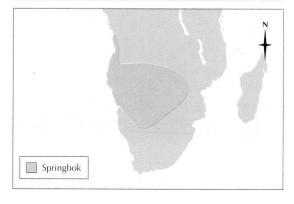

Springbok

## Migration a thing of the past

When they were abundant in the western region of southern Africa, springbok migrated in times of drought across South Africa from the Kalahari, over the Orange River, to the Cape. The migrations took the form of a massed and continuous tide of springbok. Other antelope that could not get out of the way would be engulfed and swept along. Between short, hurried bouts of feeding, the springbok pressed on relentlessly, those in front continuously being pushed onward by those behind. As a result, large-scale mishaps were common. Hundreds of animals would be drowned in the Orange River as they crossed it and, when they came to the coast, some even drowned in the sea. Because droughts occurred irregularly, so did the migrations.

The last recorded springbok migration took place in 1896. Thereafter, springbok populations did not build up again substantially enough for further mass movements. Settlers in southern Africa shot springbok for food, for sport and to prevent them from damaging crops during their extensive migrations. Their range was fenced off and divided up, producing the largely isolated, inbred populations that are still in evidence today. There are still, however, large numbers of springbok living under wild conditions in the Kalahari, particularly around Etosha Pan, a great natural oasis that has been converted into a reserve. Springbok are currently regarded as conservation dependent.

## Courtship and rearing the young

Springbok have an elaborate courtship pattern, including the *Laufschlag*, or leg-beat, similar to that of gazelles, in which the male places his stiff foreleg between the female's hind legs and makes stroking motions against her legs. This act stimulates the female to mate. The gestation lasts 168 days, at the end of which the female gives birth to one, rarely two, young.

The newborn springbok is highly vulnerable to a range of predators, including eagles and jackals. To protect her young during the first couple of days of life, the mother hides it in long grass or dense bushes.

Shortly after the birth, the females abandon the males and take the young to form nursery herds. These groups provide the individual mothers and offspring with a greater opportunity of spotting predators and also provide greater protection for the young. Young springbok start to graze when they are about 2 weeks old. By the time they are 4 weeks old they are ready to travel with the herd. They are weaned at about 4 months, although they remain dependent on their mother for a further 2 months.

*In dry periods, the springbok can obtain moisture from succulents or by pawing at the ground to reach roots. It feeds mainly at dawn and at dusk.*

# SPRINGHARE

*The springhare sits back on its hind legs when it is feeding or resting.*

THIS KANGAROO-LIKE African animal is also known as the jumping hare, or *springhaas* in Afrikaans. Its head and body grow to a maximum of about 17 inches (43 cm) and it weighs up to 8¾ pounds (4 kg). It has a relatively large, rounded head, with a blunt muzzle, prominent eyes and long, pointed, widely separated ears that tend to droop to the sides, giving the animal a rabbitlike appearance. The enlarged inner ears probably explain the springhare's sensitivity to vibration and sound. It also has long hind legs, with four toes on each. The toes have nail-like claws and the second toe from the outside on the hind foot is longer than the rest. The front legs are very short and each paw features five long, curved claws. The springhare's soft coat is sandy to reddish with some lengthy black hairs. The underparts are white, and the bushy tail is black at the end.

The single species of springhare ranges over most of Africa south of the Sahara and is also found to the northeast. Its habitat consists primarily of sandy plains and lake margins that have well-drained soils and a plentiful seasonal cover of grass. One form of springhare living in Kenya had been considered a separate species for some time, but it is now generally accepted as only a subspecies.

Zoologists generally concur that the springhare is a rodent, but there scientific agreement ends. It has variously been placed with the squirrels, the porcupines, the jerboas and the scalytailed squirrels. It is now placed in a family of its own between the scalytails and the dormice, with the name *Pedetes capensis*.

## Digs like a rabbit

The remaining information about this unusual animal is meager despite the fact that it has been known to scientists for nearly 200 years. Perhaps this is because it is small and has a primarily nocturnal lifestyle. It is never out before dark and is back underground again before dawn. The front paws are used in digging, the sharp claws serving as effective picks and rakes. The ears have a small tragus, or earlet, that may serve to keep sand out of the springhare's ears when it is digging. The springhare protects itself by closing the entrances to its burrows with plugs of earth. The burrows themselves may extend for as far as 165 feet (50 m). Warrens may spread over an area 100 yards (90 m) across. Each warren has four or five openings, and it may be inhabited by a colony, although within the warren it would appear that each individual occupies a separate sleeping chamber.

The springhare feeds on roots and bulbs, grasses, fruits, grain and similar vegetable matter, and scratches up sprouting grains to take the germinating seeds below. Insects, including locusts, are also taken from time to time.

## Nocturnal forager

There seems to be general agreement among zoologists that the springhare wanders over large areas. It moves mainly by leaping, using only the hind legs. The springhare's tail is slightly longer than its body, and helps the animal to maintain its balance when it hops.

Scientists believe that a springhare can travel as far as 6–12 miles (9.6–19.2 km) a night in search of food, and there are reports that it may wander up to 20 miles (32 km) for water during a period of drought. However, it usually travels no more than 825 feet (250 m) from its burrow. It may be able to perform leaps of 6–9 feet (1.8–3 m) on its journeys, and it is said to use roads or pathways for easier traveling. This is likely, because it does not leap very high above the

## SPRINGHARE

| | |
|---|---|
| CLASS | **Mammalia** |
| ORDER | **Rodentia** |
| FAMILY | **Pedetidae** |
| GENUS AND SPECIES | **Pedetes capensis** |

ALTERNATIVE NAME
**Jumping hare**

WEIGHT
**6½–8¾ lb. (3–4 kg)**

LENGTH
**Head and body: 14–17 in. (35–43 cm);
tail: 14¾–18¾ in. (37–47 cm)**

DISTINCTIVE FEATURES
**Small, kangaroo-like body; long, rabbitlike
ears; brown coat; white undersides; hairy
tail ending in dark brown-black brush**

DIET
**Mainly roots and bulbs; also grain, grasses,
plant stems, fruits and insects**

BREEDING
**Age at first breeding: about 2½ years;
breeding season: year-round; number of
young: usually 1; gestation period: 78–82
days; breeding interval: about 4 litters
per year**

LIFE SPAN
**Up to 19 years in captivity**

HABITAT
**Arid and semiarid habitats**

DISTRIBUTION
**Eastern and southern Africa**

STATUS
**Vulnerable**

Springhare

Springhares are safest when they are inside their burrows. Above ground they are highly vulnerable to attack from predators, including snakes, mongooses, lions and owls. They spend almost half of their time in groups of up to six when out of their burrows, probably because a group has a better chance of detecting a predator than an individual springhare.

One advantage the springhare has over many of its predators is its skill at jinking (changing direction rapidly at speed), and the secret of this ability, as with the true hares, lies in its long hind legs. However, the springhare has an added advantage over the true hares: it progresses bipedally (on two feet) and so can swerve rapidly at a sharp angle from its previous path. The rapidity of the jinking is, in fact, one of the more noticeable features of the animal's evasive tactics. If a springhare is captured, it bleats in distress and attempts to defend itself by biting its attacker and raking at it with the nails on its hind feet.

### Small litters

There is one, rarely two, young in a springhare litter and there may be up to four litters in one year. The mother is able to channel her energies into looking after only a small number of young. This reduces the strain on her and helps her to keep in a strong enough physical condition for further breeding. At birth a baby weighs just less than 9 ounces (255 g). Its eyes open at 2 days and it is soon well furred and able to move.

Young springhares emerge from their burrow only when they are fully mobile and have grown to more than half their eventual adult size. Because springhares live below ground during the most vulnerable period of their lives and only emerge when they are considerably developed, it is likely that their juvenile mortality levels are relatively low.

*The springhare forages above ground at night. It is highly vulnerable to attack at this time, particularly if there is a full moon.*

ground and appears to have difficulty in traveling through the savanna grass. A springhare can probably reach about 20 miles per hour (32 km/h) in spurts, although its average rate is probably much less than this and it cannot maintain even a reduced speed for very long.

# SPRINGTAIL

*Some species of springtails reach sexual maturity before they finish growing, and continue to molt. Pictured is a springtail of the genus Arthropleona.*

THE SPRINGTAILS ARE the most numerous and widespread primitive insects, and are equipped with a unique jumping mechanism. They are sufficiently unlike the majority of insects to make it a matter of some scientific debate as to whether they should be included in the class Insecta. The segments of the thorax, for example, are partly fused together, and the abdomen is short and made up of only five or six segments. Moreover, they have no wings, and there is no sign that they or their ancestors, found as fossils in Devonian rocks 400 million years old, ever had such appendages. However, they have three pairs of legs and agree with the more typical insects in their antennae and other features, including the simple eyes or ocelli, though these are absent in some springtail species.

Springtails are seldom more than ¼ inch (6 mm) long, and they have two unique features. The first is a fork-shaped spring called the furcula that is fastened to the hind end of the abdomen. It lies under the abdomen and is fastened with a catch known as the hamula. When released from this catch by a sudden forward movement, the fork strikes downward and throws the springtail several inches into the air. The second feature is the ventral tube on the front end of the abdomen that scientists once assumed to be a suckerlike or adhesive organ used by the springtail when climbing vegetation. This tube may also be used in regulating the springtail's water balance.

There are 6,000 to 7,000 species of springtails, with a worldwide distribution that includes the Arctic and the continent of Antarctica. Springtails are extremely diverse. Some of the species are active, dark colored, covered with hairs and scales and have long antennae. Other species are white, have short antennae, a sparse body covering and are smaller and sluggish. They also sometimes lack a jumping apparatus.

## At home everywhere

Springtails live in a wide variety of damp habitats, including soil; dead vegetation, particularly the mats that accumulate at the bases of plants; leaf litter; under fallen logs or stones and under bark. Springtails can be beneficial to humans. *Hypogastrura viatica*, for example, the natural habitat of which is the shoreline, can live in the filter beds of sewage works, feeding on algae and fungi that would choke the beds if left unchecked. Sewage workers inoculate new filter beds with a few spadefuls of material from an old one to make sure that a new population of the springtails becomes established. However, some springtails are less helpful. *Hypogastrura armata* is a pest in mushroom beds and *Tomocerus longicornis* attacks seedlings in greenhouses. *Podura aquatica* can be seen in its hundreds on the surfaces of still pools, forming masses 1 inch (2.5 cm) or more across. *Anurida maritima* forms similar groups on the surfaces of rock pools on the shore. *Achorutes nivicola* lives on the surface of snow on glaciers, in the Arctic, and temporarily when snow lies in winter in temperate latitudes. *Entomobrya mawsoni* lives under stones in penguin rookeries on the Macquarie Islands in the subantarctic. Some springtails live in ant and termite nests. *Sminthurus viridis* is round and green and so small that there may be several hundred million of them, as well as other springtails, in 1 acre (0.4 hectares) of meadow. In Australia, where it has become a serious pest, the same species is called the lucerne flea.

## Breathing through the skin

The cuticle of springtails is not waterproof as in other insects, so they breathe by taking oxygen out of water passing through the cuticle, though a small number have tracheae (air-conveying tubules). *Podura aquatica*, which lives on water, immerses from time to time but mostly stays

# SPRINGTAILS

| | |
|---|---|
| PHYLUM | **Arthropoda** |
| CLASS | **Insecta** |
| ORDER | **Collembola** |
| FAMILY | **18, including Entomobryidae** |
| GENUS | ***Sminthurides, Podura, Onychiurus, Neanura*; many others** |
| SPECIES | **About 6,000 to 7,000 species** |

**LENGTH**
**Up to ¼ in. (6 mm)**

**DISTINCTIVE FEATURES**
**Spring-loaded fork-shaped structure (furcula) on underside of abdomen, fastened to body by clasp (hamula); ventral tube**

**DIET**
**Dead or decaying organic material; fungal threads; occasionally living plants**

**BREEDING**
**Males deposit spermatophores on leaves or ground, which females collect; eggs laid in soil or protected habitats; young hatch as miniature adults; no metamorphosis**

**LIFE SPAN**
**Not known**

**HABITAT**
**Varied; includes caves, leaf litter, shorelines, vegetation and decaying plants**

**DISTRIBUTION**
**Worldwide, including Antarctica**

**STATUS**
**Abundant**

## Tiny scavengers

Springtails feed on decaying plant and animal matter or on living plants, as in the lucerne flea. Many of the litter-dwelling species reduce leaves to skeletons or feed on fungi. They probably also eat microscopic plants such as diatoms and single-celled algae. Some species have mouthparts modified for biting, whereas others are modified for sucking.

## Pincushion courting

Reproduction is simple. In many species of springtails, the males deposit their sperms in spermatophores, which are tiny rounded droplets, each on a fine stalk, like microscopic pins on the surfaces of plants or on the ground. Later, a female comes across a spermatophore and places her sexual opening over it, an act that stimulates the sperm to be released.

Growth of the young is direct; there is no metamorphosis. The hatchlings are like miniature adults. In other species of springtails, reproduction is asexual.

## Jumping insects

The spring of a springtail is unique only in structure and the way it is used. Springtails are only one of many kinds of insects that jump, such as fleas and click beetles (discussed elsewhere). The cheese maggot, the larva of a beetle, *Piophila*, is also able to jump. A cheese maggot grapples the tip of its abdomen with hooks on its mouth and then suddenly lets go, causing it to jump.

A number of insects can jump, including flea beetles, jumping plant lice, grasshoppers and locusts. Jumping insects probably evolved relatively early on in the development of species, and some scientists have suggested this habit may have led to the evolution of flight in insects. Fossil records suggest this did occur.

*A springtail, Petrobius maritimus. Some Antarctic species of springtails survive the winter by living in a form of biological stasis for many months.*

above the water because, although its claws are hydrophilic (have a strong affinity for water), the rest of its legs are hydrophobic (repel water). So the springtail rests on top of the water, held steady with only its claws in the surface film.

Some springtails emit light, in a process called bioluminescence, although there is no obvious reason for this. Many springtails have eyes comprising relatively simple light-sensitive organs called ocelli. There are six to eight ocelli on either side of the head, each made up of a ring of retinal cells with a lens above them and a cornea beyond. The larvae of higher insects also rely solely on ocelli for light reception. By contrast the compound eyes of an adult higher insect feature a collection of individual structures called ommatidia, complemented by several ocelli.

# SQUAT LOBSTER

*The squat lobster Allogalathea elegans on the arm of a crinoid. Many species sport such vivid colors but few display them, preferring to skulk in dark crevices.*

SQUAT LOBSTERS LOOK LIKE small lobsters but are more closely related to hermit crabs. The body is stout and more flattened than in a true lobster, and the broad abdomen is tucked under it. The antennae are long and slender, with a slightly stalked eye near the base of each. The long claws are strong and pointed, and the four pairs of legs all lack small pincers. Three pairs are used for walking, but the fourth pair is very small and weak and usually is tucked away out of sight.

The name squat lobster may serve a large number of species in several genera but is normally associated with the four species in the genus *Galathea*, found only in Europe. The largest of these, *G. strigosa*, is not more than 5 inches (12 cm) long in the body, which is red marked with blue lines and dots. The claws and legs and the sides of the body bear numerous spines. The smallest, *G. intermedia*, only 1 inch (2.5 cm) long including the claws, is also red marked with blue. *G. dispersa*, about 2 inches (5 cm) long, is dull red, sometimes with pale markings. Similar in size is the common squat lobster, *G. squamifera*, which is a dull greenish brown with red flecks. This is the species most often found between the tides. The others are less common and live well down the shore, usually in shallows.

## Reluctant swimmers

A squat lobster can swim as well as crawl; it darts backward by extending the abdomen and then flicking it forward in a sculling motion. Whereas a lobster, when alarmed, swims backward for long distances by flexing its abdomen, a squat lobster does not swim far but tries to crawl under the cover of rocks and pebbles. In both structure and habits it thus represents an intermediate stage between the active true lobsters and crabs and the more sedentary hermit crabs.

Squat lobsters are not difficult to find on rocky shores, especially in spring and summer, under large, flat stones. Their usual reaction on being discovered is to creep backward, trying if possible to crawl away out of sight. The largest specimens wield their claws in self-defense, but apart from this squat lobsters lack the aggressiveness characteristic of the true lobsters and crabs. Their apparent scarcity in subtidal zones could be the result of looking at the wrong time; divers at night have reported seeing large numbers in rocky areas. There are small fisheries for "squatties," which are said to taste delicious, but their small size and difficulty of processing have so far restricted commercial exploitation.

## Particulate feeders

Squat lobsters can also be said to be intermediate in regard to their feeding habits. The larger of them eat flesh in captivity, but this is unusual behavior. In the wild they do not exhibit the predatory habits typical of the larger crustaceans, preferring small particles of carrion, which are swept into the mouth by a brushing action of the bristles on the mouthparts.

## Primitive larvae

The life cycle of squat lobsters is similar to that of other large crustaceans, but the zoea larvae are more primitive and less highly specialized in structure than those of true crabs. They are more like those of shrimps and prawns. At a later stage in the larval life, they look like tiny but incomplete shrimps. The larvae of even the largest squat lobsters are only ¹⁄₁₆ inch (1.5 mm) long.

## Out on a limb

The squat lobsters are important for the light they shed on the relationships of the decapod (10-legged) crustaceans. On the one hand are the shrimps, prawns and lobsters, with the long abdomen normally extended backward. On the other hand are the true crabs, with the abdomen small and tucked in under the rest of

# SQUAT LOBSTER

| | |
|---|---|
| PHYLUM | **Arthropoda** |
| CLASS | **Crustacea** |
| ORDER | **Decapoda** |
| FAMILY | **Galatheidae** |
| GENUS AND SPECIES | ***Galathea strigosa*** |

LENGTH
**Up to 5 in. (12 cm); usually 3½ in. (9 cm)**

DISTINCTIVE FEATURES
**Bright red body with vivid blue stripes and blotches; otherwise resembles a shortened lobster with particularly hairy legs**

DIET
**Carrion and small organic particles**

BREEDING
**Eggs brooded on female's underside for a few weeks before hatching; larvae spend a few days or weeks among animal plankton before settling; complex series of larval and postlarval stages before adult form reached**

LIFE SPAN
**Probably 5–10 years**

HABITAT
**Gravel and rock seabeds from shallow shore waters to depths of 650 yd. (600 m) or so; most numerous where there are plenty of crevices in which to hide**

DISTRIBUTION
**Eastern North Atlantic, from Norway south to Mediterranean; also Red Sea**

STATUS
**Locally common**

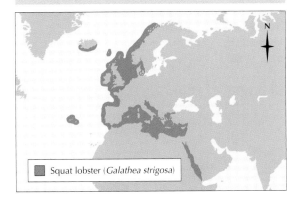

Squat lobster (*Galathea strigosa*)

*Galathea strigosa, of the Mediterranean Sea, is the largest species of its genus. Other squat lobster genera include* Munida *and* Munidopsis.

the body. In between are the hermit crabs, with an asymmetrical abdomen and the legs much reduced, those on one side of the abdomen being absent. At first sight the hermit crabs, despite obvious affinities with shrimps, prawns, lobsters and crabs, have diverged from the standard body plan, and along the way have specialized in sheltering in empty shells.

If naturalists were to assemble all the European squat lobsters and their related neighbor, they could build up a series illustrating how this body plan evolved. The abdomen of squat lobsters is slightly asymmetrical. That of the porcelain crab, *Porcellana longicornis*, often found under stones in company with squat lobsters, has the abdomen slightly more asymmetrical. Another relative nearby, the burrowing prawn, *Callianassa subterranea*, looks like a normal prawn but lives in a burrow in the sand; like the hermit crabs, it has one claw much larger than the other. The stone crab, *Lithodes maia*, is spiny like the squat lobsters and looks like a spider crab, which is a true crab but has a markedly asymmetrical abdomen.

The story is repeated, with local variations, on the North American coasts. For example, there are the porcelain crabs *Pachycheles* and *Petrolisthes*, almost indistinguishable except to the expert eye from the European porcelain crab. There are also stone crabs (genus *Lopholithodes*), sand crabs (genus *Emerita*) and mole crabs (genus *Blepharipoda*), so named because they burrow. In these, also, naturalists could trace the tendency toward asymmetry; the tendency to become particulate feeders rather than predators; and the tendency to shelter in crevices, cracks or burrows, that have taken the hermit crab along its own evolutionary path.

# SQUID

**Squid, like all cephalopods, have huge eyes, large optic nerves and superb color vision for use in prey capture and communication.**

SQUID ARE MARINE MOLLUSKS of the class Cephalopoda, which also includes the octopuses, cuttlefish and nautiloids. Unlike their relatives they have a streamlined body and are fast swimmers. There are about 350 species, ranging in size from less than 1 inch (2.5 cm) to nearly 60 feet (18 m). Squid are found in all oceans, from the poles to the Tropics, and from the shallows to the depths.

A squid's head bears two well-developed eyes, very like human eyes in structure although independently evolved; eight arms corresponding to those of an octopus; and two longer tentacles. The body tapers to the rear, bears lateral fins at the back and contains the remains of the ancestral shell. The underside of the body contains the mantle cavity into which open the reproductive and excretory organs, the ink sac and the hind end of the digestive tract. Here also are the two gills in which the squid's blue blood is oxygenated. The pulsations of the muscular wall of the mantle cavity continually draw water in and out through a muscular funnel that opens behind the head. Like other cephalopods, the squid uses this water current for jet propulsion, expelling it through the funnel, which can be directed forward or back to send the squid rapidly in either direction. The water is sucked in, not through the funnel, but around its base.

That this system works blindingly fast is due to highly efficient nerve fibers linking brain and muscle. The nerve fibers of mollusks lack the covering that facilitates the very rapid conduction of nerve impulses in mammals. However, the rate at which the fibers can conduct impulses increases with their thickness. Squid have evolved very thick nerve fibers to link brain and mantle muscle and ensure that movements are synchronized to the split second. They are over 0.5 millimeters thick, as compared with diameters of one-thousandth to one-fiftieth of a millimeter in humans. Study of these giant fibers has been of immense help in exploring how nerve impulses are conducted through the body of an animal.

## Illuminated squid

Squid tend to move about in shoals, often of only one sex. They swim well in either direction but travel mostly backward. Like the octopus, they are masters of color change and have chromatophores (pigment cells) of various colors in the skin that can be contracted and expanded to produce a variety of patterns. *Sthenoteuthis*, for example, has chromatophores of the three primary colors—blue, yellow and red.

One of the best-known species is *Loligo forbesi* of the North Sea, the northeastern Atlantic and the Mediterranean. Large individuals may be

# SQUID

PHYLUM **Mollusca**

CLASS **Cephalopoda**

ORDER **Teuthoidea**

FAMILY **Architeuthidae, Histioteuthidae, Chiroteuthidae and Loliginidae**

GENUS AND SPECIES **Many, including European squid, *Loligo forbesi*, detailed below**

LENGTH
**Up to 24 in. (60 cm), or 39 in. (1 m) including longest tentacles; usually smaller**

DISTINCTIVE FEATURES
**Medium-sized squid with angular, lozenge-shaped fins on posterior end**

DIET
**Small fish and crustaceans**

BREEDING
**Male passes packets of spermatophores into female using modified arm; female lays strings of eggs on seabed attached to stones or seaweed; young hatch out as miniature adults; parents die after breeding**

LIFE SPAN
**Rarely more than 1 year**

HABITAT
**Coastal and offshore waters, near seabed or in midwater**

DISTRIBUTION
**Eastern North Atlantic and Mediterranean**

STATUS
**Common**

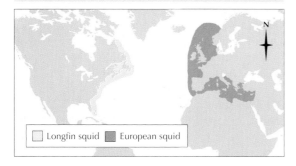

Longfin squid    European squid

2–3 feet (60–90 cm) long, but most are 8–12 inches (20–30 cm). Less typical in appearance is the small *Chiroteuthis veranyi*, with its tentacles many times the length of the body, and the reddish-purple to black *Histioteuthis*, with six of its arms joined by a membranous web whose jellyfish-like movements largely replace the use of the funnel in swimming. In *Bathothauma*, a small deep-sea squid, the very short arms are carried in a small rosette around the mouth on the tip of a long peduncle (stalk); the eyes, too, are on long stalks. *Calliteuthis* also has unusual eyes in that one is much larger than the other. The smaller eye is surrounded by a circlet of light-producing organs.

Many squid have light organs, especially those living in deep water. The light organs may be on the eyes, on the tentacles or scattered all over the body. *Histioteuthis bonelliana*, a Mediterranean species about a yard (90 cm) long, has nearly 200 of them. The light organs themselves may be highly complex, with reflectors, lenses, diaphragms, color screens and other accessory structures. Some squid do not themselves produce the light but have luminescent bacteria in their light organs that are transmitted to the next generation in the squid's embryo. The light is sometimes given out from internal organs and shines through a translucent body wall.

## Leviathans of the deep

The largest species of squid belong to the genus *Architeuthis*, in particular the species *A. dux*. A specimen of *A. princeps*, stranded on the coast of Newfoundland in 1878, had a body 20 feet (6 m) in length and a total length of 55 feet (16.7 m). This was exceeded only by a specimen of the less bulky *A. longimanus*, 57¼ feet (17.5 m) long, found in Lyall Bay, New Zealand, in 1888. Smaller specimens of *Architeuthis* have been stranded on the shores of the British Isles. These giants, some of which may exceed even the recorded maximum sizes, probably lie behind tales of the kraken, a dreaded monster, in Norse legend. From their anatomy they are thought not to be powerful swimmers and are believed to live at a depth of 600–1,200 feet (180–365 m). Giant squid are known to prey on whales. In

*Pop-eyed squid, Sepioteuthis sepioidea, mate and lay their eggs by night off the Bahamas. The male uses a specialized arm, called the hectocotylus, to transfer sperm packages to his mate.*

*Smaller squid prey on crustaceans and other marine invertebrates that migrate daily through the water column: rising by night, diving again by day.*

The eggs are laid in strings of jellylike material that become attached to the seabed. *Loligo* dies after spawning. Courtship is a communal activity in squid, and great shoals may gather for the purpose, sometimes then migrating inshore to breed. The number of eggs produced by a shoal may be colossal, although the egg masses of 40 feet (12 m) or more across that were reported off California in 1953 were exceptional. In 1955 it was estimated that in this locality the eggs covered 200 acres (80 ha) of seabed. The young hatch as recognizable squidlike forms.

## Laying a smoke screen

Squid are preyed on by albatrosses, petrels and other seabirds, and they are the main food of the king penguin, *Aptenodytes patagonica*. The emperor penguin, *A. forsteri*, also eats them in large numbers, as do seals, sea lions, elephant seals, toothed whales and tuna. For escape squid rely partly on their agility and excellent vision, but also on their ability to change color and to blow ink out through the funnel. The ink may be used simply as a smoke screen, but in some species it remains for a while as a compact cloud in the water, approximately the size and shape of the squid producing it. Leaving this decoy, the squid darts off backward, changing to a different and less conspicuous color as it does so. It is believed that the predator attacks the blob of ink and loses sight of the fleeing squid.

Squid are fished commercially for food, especially in Antarctic and subantarctic waters. They are caught on hook and line, on which special lures jiggle up and down to attract them.

## Flying squid

There are at least two species of flying squid. One of these, *Onychoteuthis banksi*, occurs in all oceans and is sometimes stranded on the shore. Its fins are wide, and further planes are provided by broad membranes on the arms. It was the account of Thor Heyerdahl's *Kon-Tiki* expedition from Peru to Polynesia in 1947 that brought flying squid to general attention, although they had long been known to sailors because they not uncommonly landed on the decks of ships. These extraordinary animals are hooked squid that can leave the water with such velocity that they may sail 50–60 yards (45–55 m) through the air before reentering the sea. They sometimes leap singly, sometimes in twos or threes. They have been known to hit ships as high as 20 feet (6 m) above the waterline. The flying tactic is probably a means of escape from predators.

1966 observers on the coast of South Africa saw a giant squid take a baby southern right whale, *Eubalaena australis*, from its mother. A tougher adversary is the sperm whale, *Physeter catodon*, adults of which have been found with remains of giant squid in their stomachs. Sucker-shaped weals on the whale's skin testify to titanic deepsea struggles between mammal and cephalopod.

## Venomous bite

Squid feed mainly on fish and crustaceans, as well as on smaller squid. Having seized prey, they quickly paralyze it using venom produced by one of the two pairs of salivary glands. *Loligo* seizes a fish behind the head and bites it off with its parrotlike beak; the rest is bitten into small pieces and digestion is completed in 4–6 hours. Squid may chase their prey or make use of their camouflaging abilities to ambush it.

## Massive matings

In temperate seas squid usually breed in spring or summer. In *Loligo peali* of American waters, which is typical, the male initiates courtship by displaying to a female with his arms. Colored spots flare on his arms, and from time to time he blushes red all over his body. He then lies parallel to his mate, lower side to lower side and with his arms around her, just behind the head. His sperms are packed into spermatophores, torpedo-shaped tubes of chitin about ⅗ inch (15 mm) long. He transfers bunches of these elaborate structures from his mantle cavity to the female by means of his left fourth arm, which is specially modified for the purpose. Alternatively, the pair may join head to head, in which case the sperm packets are deposited on a glandular patch among the female's tentacles.

# SQUIRRELFISH

SQUIRRELFISH ARE USUALLY bright red and have large eyes, and this faint resemblance to the red squirrel seems to be the only reason for their common name. They have a deep body, a large head, strong jaws and prominent eyes. The largest species, the sabre squirrelfish (*Sargocentron spiniferum*), exceeds 2 feet (60 cm) in length, but 1 foot (30 cm) is more usual.

The scales covering a squirrelfish are large, with sharp points on their hind edges. There are sharp spines on the head and on the gill covers, and it should be noted that these fish are very prickly to handle. The front half of the dorsal fin is spiny, whereas the rear half is tall and soft-rayed. The anal fin has four spines in front. The large pelvic fins are situated forward on the body and level with the pectoral fins. The tail fin is forked. The red body usually sports silvery spots or stripes on the flanks, from behind the gills to the base of the tail fin.

There are 80 species of squirrelfish in tropical seas throughout the world, most of them living in shallow water. A widespread species in the Indian and South Pacific Oceans is the redcoat, *S. rubrum*, which lives in deeper water than most, down to 90 feet (30 m). It is bright red, and each row of scales along its body bears a silvery stripe. Its fins are rosy, with black and silver markings.

## Unneighborly fish

Most squirrelfish are active at night. By day they shelter singly in crevices and cracks in the coral. Each occupies and defends a territory. Some members of a related genus, *Myripristis*, contrast with typical squirrelfish in sometimes forming shoals.

Squirrelfish are noted for the noises they make, which are loud enough to be heard above the water. These are produced by the vibration of muscles attached to the swim bladder, which acts as a resonator. The sounds are said to serve the same basic purpose as bird song: to stake claim to a territory and to bring pairs together for breeding.

Squirrelfish are predators that hunt smaller fish, and one benefit of their pronounced territorial instinct is that it keeps the individuals well spaced out, minimizing competition with neighbors. This probably is when they use sound to warn possible trespassers off their beat. Fishers used to exploit this territorial instinct off the Hawaiian Islands, where squirrelfish have in the past been an important food resource. A single squirrelfish was caught in a net; a string was tied around the live fish, which was returned to the water and dangled near crevices in the rocks. Other squirrelfish soon emerged to fight it, and by drawing the captive fish gently to the surface, the fisher caught the rest by easing a net below them.

## Helpless larvae

The connection between the sounds they make and their breeding habits was first discovered by accident, among squirrelfish in a display tank in a US television studio. The sounds were heard during a rehearsal, and a pair of squirrelfish was seen to be courting, the fish lying side by side with their tails pressed together and their bodies forming a V.

## Hazards of infancy

The larvae that hatch from squirrelfish eggs are remarkable for their long, pointed snouts. After emerging, they swim to the surface and drift on the currents with the rest of the animal plankton. This is the most perilous stage in a squirrelfish's life, when the vast majority of larvae fall prey to tuna and other large fish. As adults, with their nocturnal habits, spininess and tendency to stay hidden, squirrelfish have relatively few enemies, other than humans, who catch them for food.

*The candy-striped coloration of this fish identifies it as the redcoat, a deepwater species of the Indian and South Pacific Oceans.*

*Large eyes give the squirrelfish an owl-like appearance. This is a longspine squirrelfish,* Holocentrus rufus, *in the Caribbean Sea.*

## SQUIRRELFISH

| | |
|---|---|
| CLASS | **Osteichthyes** |
| ORDER | **Beryciformes** |
| FAMILY | **Holocentridae** |

GENUS AND SPECIES **80 species, including redcoat, *Sargocentron rubrum*, and sabre squirrelfish, *S. spiniferum* (detailed below)**

LENGTH
**Up to 24½ in. (61 cm)**

DISTINCTIVE FEATURES
**Fins orange or vermilion; distinguished from violet squirrelfish, *S. violaceum*, by absence of white tips to dorsal fin; venomous spines on gill covers**

DIET
**Crabs, shrimps and small fish**

BREEDING
**Larvae develop among animal plankton**

LIFE SPAN
**Not known**

HABITAT
**Variety of reef zones, from flats to lagoon and seaward reefs, to depths of at least 135 yd. (122 m); hides under ledges by day**

DISTRIBUTION
**Indian and South Pacific Oceans: Red Sea and East Africa to Hawaiian and Ducie Islands, north to Japan, south to Australia, throughout Micronesia**

STATUS
**Relatively common in areas not subjected to heavy spearfishing**

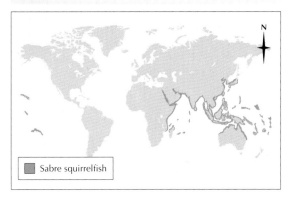

Sabre squirrelfish

## Lights to their eyes

Squirrelfish are said to be primitive, a reference to certain details of their anatomy. They form a link between the multitude of perchlike fish living today and certain fish, dominant during the Cretaceous period 135–70 million years ago, which had spiny rays on their fins. More primitive still are the alfonsinos, family Berycidae. These too are brightly colored, but differ in having a rounded body with a long, slender tail stock. They also differ in that they inhabit deep water, at about 2,000 feet (600 m). A common species, the 2-foot (60-cm) long splendid alfonsino (*Beryx splendens*), is found worldwide in warm seas and is fished commercially.

Another relative of the squirrelfish is the pinecone fish, *Monocentris japonica*, of the family Monocentridae. A few inches long, it has platelike, spiny scales and lives in deep water in tropical parts of the Indian and South Pacific Oceans. It is eaten in Japan. This family also contains the pineapplefish, *Cleidopus gloriamaris*, of Australian waters. It has large, almost wistful eyes, and under its lower jaw pockets are two light organs filled with luminescent bacteria.

Strangest of all are the related lantern-eyed fish of the family Anomalopidae. None is longer than 1 foot (30 cm). They have a light organ beneath each eye made up of tubes of luminous bacteria. The fish cannot control the bacteria but they can cover the light; some species draw a lid over the organ, while in others a muscle twists the organ so its light is no longer visible from outside. The light serves to attract animal plankton toward the fish to be eaten.

# SQUIRREL MONKEY

THE SMALL, SLENDER squirrel monkey is known in German as the *Totenkopfaffe*, or the death's-head monkey. This name is inspired by the the skull-like appearance of its small white face and its large dark eyes. It has long, thin hind legs, rather shorter arms and a short body. The head is rounded, with a small face and a slightly protruding black muzzle. The nostrils, as in other monkeys with a South American distribution, face sideways. The ears are rather pointed, with long white hairs directed out to the side.

The tail is not prehensile, but tends to be held in a coiled position and often partly curls around something as a support. The tail is 13¾–15¾ inches (35–40 cm) long; the head and body are only 10–12½ inches (26–32 cm) long.

## Variation in color

Squirrel monkeys have short, thick fur. The muzzle is black and naked, and the fur around the eyes and on the cheeks is white. The head is blue gray, the upper body is yellowish with gray undertones and the underparts are whitish cream. The forearms and tail tip are black.

There are five species of squirrel monkeys: *Saimiri boliviensis*, *S. vanzolinii*, *S. sciureus*, *S. ustus* and *S. oerstedii*. They vary considerably in color, according to their geographic location. The Central American form has a jet black cap and the yellowish color of the body is overlaid with red. Two South American squirrel monkeys, from Brazil and Peru respectively, are referred to as the Gothic and Roman squirrel monkeys because the white arches above the eyes are said to recall the architectural styles and features of those periods.

## Latin American distribution

In Central America, squirrel monkeys are found along the coastal strip on the Pacific side of the Isthmus of Panama. The same race extends down the Pacific coast of South America as far south as Ecuador. To the east of the Andes, they inhabit the tropical forest belt and much of the subtropical forest, extending from the Orinoco River, south across the Amazon, to the Mato Grosso region of Brazil, and possibly past the Llanos de Guarayos. Squirrel monkeys can be found in Peru, Bolivia, Columbia, Guyana, Panama and Costa Rica.

## Loosely organized troops

Squirrel monkeys live in very large bands, or troops, kept together by the females, which lead the troop's movements and form a focus for both young and adult males. The males keep to the

*Squirrel monkeys are slim, highly active monkeys that form some of the largest bands of any primate.*

## SQUIRREL MONKEYS

| | |
|---|---|
| CLASS | **Mammalia** |
| ORDER | **Primates** |
| FAMILY | **Cebidae** |

GENUS AND SPECIES    **5 species, including
*Saimiri sciureus, S. boliviensis* and *S. ustus***

ALTERNATIVE NAMES
**Titi; mono**

WEIGHT
**Female: 23 oz. (650 g); male: 33½ oz (950 g)**

LENGTH
**Head and body: 10–12½ in. (26–32 cm);
tail: 13¾–15¾ in. (35–40 cm)**

DISTINCTIVE FEATURES
**Short, thick fur; black, naked muzzle; white
eyes and cheeks; blue-gray head; yellowish
gray upper body; white-cream underparts;
black forearms and tail tip**

DIET
**Fruits, insects, leaves and seeds**

BREEDING
**Age at first breeding: 2½–3 years (female),
4–5 years (male); breeding season:
February–April or May; gestation period:
147–170 days; number of young: 1;
breeding interval: 1 year**

LIFE SPAN
**Up to 25 years in captivity**

HABITAT
**Intermediate canopy of tropical dry forest
and rain forest**

DISTRIBUTION
**Panama and Costa Rica south to Brazil,
Peru, Bolivia, Columbia and Guyana**

STATUS
***S. oerstedii:* endangered; *S. vanzolinii:*
vulnerable; other species locally common**

*The squirrel monkey's
tail, often longer than
its head and body, is
used as a support and
balance, particularly
when feeding.*

edges of the troop, having little to do with each
other or the rest of the troop, except in the
mating season. Mutual grooming is very infrequent, and although some females are more
influential than others, there is no marked female
hierarchy. There is more hierarchy among male
squirrel monkeys, but because of the low level of
social organization this is not really apparent
except in the breeding season.

Squirrel monkeys inhabit deep jungle and
forest edge habitats as well as gallery forest,
bordering large rivers. They are extremely versatile ecologically, coming down to the ground and
going up to the very tops of the trees. Their food
consists of fruit, buds and insects, with some leaf
matter. The squirrel monkeys wrap their tails
around branches when feeding, in order to use

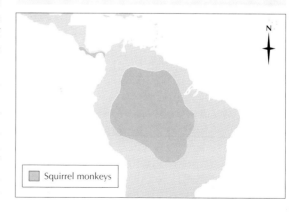

Squirrel monkeys

them as stabilizers. Squirrel monkeys wrap their tails around the body when they are resting. The monkeys often rest in groups, huddled together with their heads between their knees.

## Each troop a matriarchy

The breeding season is not sharply defined, but is a period of about 4 months, from February to about May, during which both sexes undergo sexual development. There is a birth peak near the end of the season. The behavior of all the male squirrel monkeys changes completely during the breeding season, as does their appearance. They become much fatter in the arms, chest and head. They also become very vocal and highly aggressive, displaying to one another and chasing the females to mate with them. Mating reaches a peak in the dry season, and the young are born 147–170 days later, at the height of the rainy season. When a female is giving birth, the other, mainly female, squirrel monkeys gather around her and closely observe the proceedings.

The newborn squirrel monkey may be carried around by either male or female, but the male's interest is short-lived and the only thing that keeps him near the troop is the presence of the females. The young can grip its mother's fur and stay clinging to her even when it is asleep. Squirrel monkeys breed well in captivity and have been known to live for up to 25 years.

## Easy prey

Because of its small size, the squirrel monkey falls easy prey to small cats, birds of prey, pythons and humans. Its only defense is to run away. When one monkey gives an alarm call, the whole twittering troop falls silent. The males go to investigate the source of the disturbance and then return and lead the troop in the opposite direction, away from any impending danger.

## Monkey language

Scientists have made many studies of squirrel monkeys' behavior. One group of scientists has concentrated on studying their use of sounds. They identified 26 separate calls, made up of different combinations of a few basic sounds, including peeping, twittering and shrieking. Human language is also built up of basic elements, the number differing from one language to another, and so these studies provide one more link between monkeys and humans. The vocalizations differ slightly between the Gothic and Roman squirrel monkeys.

*Squirrel monkeys live in the intermediate canopy levels of tropical dry and rain forests. Like all South American monkeys, they have sideways-facing nostrils.*

# STARFISH

least a thin covering of skin, although
this may wear through in places. They
may occur as closely set plates or may
form an open network. Spines of the
same material project from the surface
singly or in groups, each spine moved
by muscles at its base.

*A scarlet nubbly starfish, Oreaster occidentalis, off the Galapagos Islands in the Pacific Ocean. The network of ossicles (tough platelets) can be clearly seen.*

NO ANIMAL IS MORE clearly symbolic of the sea than the starfish or sea star, which occurs in one form or another in all the world's marine waters. There are 1,600 species, most of which live in shallow waters, although some live in deep seas. Species are most numerous at high latitudes of the North Pacific. Starfish are members of the Echinodermata, the large phylum of marine invertebrates that also includes brittle-stars, sea urchins, sea cucumbers and sea lilies.

## Radial symmetry

The typical starfish form is made up of five arms radiating from a small central body with a toothless mouth on its underside, but there is plenty of scope for variety. The number of arms may range from four to 50, and some of the common starfish that normally have five arms may have from three to seven. The smallest species are less than ½ inch (12 mm) across, whereas the largest are 3 feet (90 cm). Common colors are yellow, orange, pink and red, but some starfish are gray, blue, green or purple. Some of the smallest, known as starlets or cushion stars, have such short arms that their outline is effectively pentagonal.

The body wall of a starfish is reinforced and supported by calcareous plates, or ossicles, more or less exposed at the surface but always with at least a thin covering of skin, although this may wear through in places. They may occur as closely set plates or may form an open network. Spines of the same material project from the surface singly or in groups, each spine moved by muscles at its base.

The surface may also bear many little pincerlike pedicellariae, like those of sea urchins. The pedicellariae take various forms. Some are a pair of tiny jaws mounted on a short stalk, whereas others consist simply of three spines with their bases close together. They play a key role, seizing small organisms and thus preventing the surface from becoming encrusted with algae and sedentary animals. The pedicellariae are aided by cilia (beating, hairlike organs) on the surface. The material of the ossicles, spines and pedicellariae of echinoderms is unique in the animal kingdom, in that each element is a single crystal of calcite growing in the form of a three-dimensional network, combining lightness with strength.

In addition to a general covering of sensory cells, there is a light-sensitive optic cushion at the base of a short tentacle, which is a modified tube foot at the tip of each arm.

## Walking on water

Starfish move about by means of numerous tube feet arranged in two or four rows along a groove on the underside of each arm. The tube feet are hollow, muscular cylinders connecting at their bases with the water vascular system, a system of tubes that are filled with water. The tube feet are pushed out hydraulically by the contraction of muscular sacs that lie at intervals along the vascular system. At their tips there usually are suction discs, which also have sticky secretions that aid them in sticking to rock or prey. In the burrowing starfish the tube feet lack suckers.

The water vascular system connecting the tube feet opens to the outside through one or more porous plates on the upper surface. These madreporites, usually single, are situated off center of the body disc.

## One arm leads the way

In some species one arm nearly always takes the lead when the starfish is walking, but it is more usual for the arms to take turns leading the way. However, there are variations in the extent to

## BURROWING STAR

| | |
|---|---|
| PHYLUM | **Echinodermata** |
| CLASS | **Asteroidea** |
| ORDER | **Phanerozonia** |
| FAMILY | **Asteropectinidae** |
| GENUS AND SPECIES | *Astropecten irregularis* |

**ALTERNATIVE NAME**
**Sand star**

**LENGTH**
**Across arms: 4–8 in. (10–20 cm)**

**DISTINCTIVE FEATURES**
**Rigid, star-shaped body with 5 tapering arms and mouth on underside; sandy yellow to pale violet coloration; large, square plates and long, pointed spines on edge of each arm; tube feet in deep groove extending from mouth to tip of each limb**

**DIET**
**Mollusks, worms, crustaceans, fish and other echinoderms; also scavenges**

**BREEDING**
**Dioecious: sperm and eggs (2 to 3 million) shed into water, where fertilization occurs. Egg hatches into bipinnaria larva, then metamorphoses into brachiolaria larva with 3 arms; larva develops for around 2 months before setting on seabed as tiny starfish.**

**LIFE SPAN**
**Not known**

**HABITAT**
**Seabed at depths of 30–3,000 ft. (10–1,000 m), lying partly buried in clean sand or sandy mud**

**DISTRIBUTION**
**Atlantic, from Norway to Morocco, including British coasts; Mediterranean**

**STATUS**
**Very common**

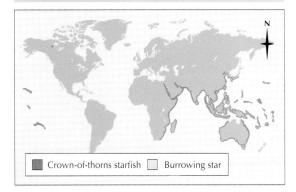

Crown-of-thorns starfish    Burrowing star

which each arm is favored in a given species or individual. One species has been known to travel at the relatively breakneck speed of 2 yards (1.8 m) a minute, but a more usual speed is 2–3 inches (5–7.5 cm) a minute.

## Multiplying by dividing

Irregular starfish are sometimes found: those that have lost one or more arms and are regenerating new ones. Their impressive powers of regeneration have been observed on oyster and mussel beds, where they are frequently pests. Those whose occupation it was to dredge the starfish and get rid of them used to tear them apart and throw them back. This proved to be a waste of time: the damaged starfish were able to regrow from the torn parts. In most species a part of the body is needed for regeneration, but one genus, *Linckia*, is known to be able to regenerate from just a piece of arm ½ inch (12 mm) long. *Linckia* actually uses its arms to propagate itself, the limbs pulling in two directions until the animal tears itself in two. Any bits that break off add to the number of new individuals.

## Eating out

Most starfish are predators, feeding on mollusks, worms, crustacea, fish and other echinoderms. Those, like *Asterias,* that prey on bivalves open them by arching over them and pulling on the shell valves with their tube feet. The mollusk may resist, but the starfish eventually overwhelms it and the bivalve, as the result of muscle fatigue, has to allow its valves to part a little. The starfish then protrudes its stomach and inserts it, inside out, into the mollusk—a slit of 0.1 millimeter is enough for it to make an entry. The stomach then secretes digestive enzymes into the mollusk.

*The female blood sea star, or bloody Henry (seen here in waters off Cornwall, England), broods her eggs until they hatch by arching over them.*

The burrowing or sand star, *Astropecten*, feeds differently, by taking food in whole. Shells or skeletons are later ejected through the mouth, because this genus has no anus. The cushion star, *Porania pulvillus*, sometimes cast up on the beaches of Europe, is unusual in that it feeds on microscopic organisms, propelling them toward its mouth by means of the cilia on its underside. Another species, *Ctenodiscus crispatus*, feeds on mud drawn into its mouth in strings of mucus along the grooves under the arms. *Asterina gibbosa*, one of the cushion stars, eats sponges and sea squirts.

## Born in a stomach

Regardless of its gender, a starfish has two reproductive organs in each arm, each opening by a pore at the base of the arm. There usually is one breeding season in a year, when millions of eggs may be released into the sea. *Asterias* may release 2–3 million within two hours, but as many as 200 million are released by some species.

In *Asterias*, a bipinnaria larva hatches from each egg. It has two circlets of cilia and is bilaterally symmetrical. The fore end later becomes drawn out into three arms, the larva then being called a brachiolaria. A curious asymmetrical

development results in the growth of a young starfish mainly from the left side of the larva, of which it still remains part.

After about two months spent drifting on currents with other plankton, the larva anchors itself by its three adhesive arms and the young starfish breaks free from the rest. Some cushion stars attach their eggs to the undersides of stones, the brachiolaria stage being omitted. The change into a starfish therefore occurs at an earlier stage of larval development. Several species brood their eggs, and these hatch as young stars rather than as larvae. In these species, which mostly live in colder waters and particularly in the Antarctic, the eggs are large and yolky and less numerous. In some species, such as the blood sea star, *Henricia sanguinolenta*, the mother arches herself over her sticky eggs until they hatch. Meanwhile she goes without food. Among other odd methods of brooding is that of *Leptasterias groenlandica*, in which the eggs are kept in pouches in a parent's stomach.

## Incredible story

The raids by starfish on oyster and mussel beds are insignificant compared with the large-scale destruction of coral reefs that took place some years ago and seriously affected fisheries, with the added danger of land erosion. The arch villain is the crown-of-thorns starfish, *Acanthaster planci*, so named for its covering of spines. It has 16 arms and averages 10 inches (25 cm) across, although it can reach 2 feet (60 cm). It feeds on coral polyps.

The crown-of-thorns was thought a rarity until about 1963, when swarms were reported on the Great Barrier Reef. At the same time, it was implicated in the destruction of coral in the Red Sea. Populations exploded in many widely separated areas of the Pacific and other oceans, killing off coral at an alarming rate. In 2½ years the species killed nine-tenths of the coral along 24 miles (38 km) of the coast of Guam. As the polyps are destroyed, the dead coral is overgrown with weed and most of the fish depart, their habitat ruined. The areas affected include the Great Barrier Reef of Australia, Fiji and Palau. At first it was thought the population explosion among crown-of-thorns was due to pollution or other human interference, such as overcollection of the starfish's predators, including the giant triton shell, *Charonia tritonis*, for the seashell trade. To date, though, no satisfactory explanation has been offered.

*A crown-of-thorns starfish wraps itself around a coral head as it everts its stomach to digest the living polyps. Swarming periodically, these starfish have wrought havoc on tropical reef systems.*

# STARGAZER

THE STARGAZER IS ABOUT as well equipped as any animal for getting an easy living and beating off attacks on itself. To humans it also is one of the ugliest of all fish. Its head is large, broad and flat on top, and the body, which is covered with small scales, tapers evenly from the head to the square-ended tail fin. Stargazers rarely exceed 1 foot (30 cm) in length, the largest being the northern stargazer, *Astroscopus guttatus*, of the Atlantic coast of the United States, which reaches 22 inches (56 cm).

The mouth is wide and the jaws are set almost vertically in bulldog fashion. The wide-set eyes on the top of the head gaze permanently upward, and behind each eye is a rhomboidal depression in the skin marking the location of electrical organs. The front dorsal fin is short and spiny, the second dorsal is long and soft-rayed and the anal fin is also long and soft. The gill covers are large, the pectoral fins are large and set low on the body and the pelvic fins are small and set under the throat. There are venom spines just above the pectoral fins, each with two grooves that carry poison from a gland at their base. The dull brown of the body may be broken by whitish spots or stripes. In all cases the colors make the fish inconspicuous on any seabed. There are eight genera containing a total of 45 species, occurring mainly in tropical and subtropical seas throughout the world.

## Everything to protect itself

Stargazers bury themselves in sand or mud by a squirming side-to-side motion in which the large pectoral fins seem to act as shovels. Once buried, they move about very little, lying with just the eyes and nostrils showing above the surface; if need be, they can bury themselves temporarily to a depth of 1 foot (30 cm). In most stargazers the nostrils open into the mouth, which is unusual in fish, and water is drawn in through them to pass across the gills. The lips are fringed with short fleshy tentacles that may act as a filter to keep out sand.

The electrical organs can generate 50 volts, enough to cause anyone who touches a stargazer to throw the fish aside in surprise. They are formed from modified eye muscles, each of the electroplates representing a single muscle fiber. The venom spines are a second form of defense.

*A species of stargazer on shale in Sulawesi, Indonesia. The eyes protrude a little from the top of the head. When the fish has partly buried itself in the seabed, they still give a field of view.*

*The broad, scalloped pectoral fins serve as shovels, enabling the stargazer to wriggle into the seabed and adopt a low profile, from which it can ambush its prey.*

## STARGAZERS

| | |
|---|---|
| CLASS | **Osteichthyes** |
| ORDER | **Perciformes** |
| FAMILY | **Uranoscopidae** |

GENUS AND SPECIES **8 genera containing 45 species in total, including Atlantic stargazer, *Uranoscopus scaber* (detailed below)**

LENGTH
**Up to 16 in. (40 cm)**

DISTINCTIVE FEATURES
**Elongated and rounded body; laterally flattened head with rough, bony plates; very small eyes set on top of head; large, almost vertical mouth; feeding lure on lower jaw; dull brown body, pale below and dark above, may have white markings; electrical organs behind eyes**

DIET
**Mostly fish; also shrimps and worms**

BREEDING
**Breeding season: April–August; planktonic eggs and larvae**

LIFE SPAN
**Not known**

HABITAT
**Sandy or muddy seabeds**

DISTRIBUTION
**Eastern Atlantic, from Spain south to Senegal; also Mediterranean**

STATUS
**Not threatened**

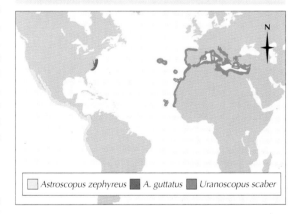

☐ *Astroscopus zephyreus* ■ *A. guttatus* ■ *Uranoscopus scaber*

## Attracting prey with a lure

When a stargazer lies buried in the sand, its mouth is more or less level with the surface of the seabed, and any unwary little fish that swims within reach is soon sucked down and snapped up. Other prey includes small crustaceans and worms. The stargazer also has a fleshy, wormlike filament rooted just inside the mouth, under the tongue, that can be pushed out and waggled to arouse the curiosity of larger fish, drawing them nearer the mouth.

## Normal infancy

Although stargazers do not normally move about much, there seems to be some migration, probably linked with temperature. Stargazers sometimes move into shallow temperate seas in summer, for example. They also appear to swim into deeper waters to spawn. Until they are about 1 inch (2.5 cm) long young stargazers resemble most other young perchlike fish, with the eyes at the sides of the head and mouth horizontal. Then, as the head flattens and the eyes move up, certain muscles of the eye change to form the electroplates.

## Looking at the stars

Stargazers sometimes appear under the scientific name *Astroscopus*, meaning "to look at the stars." This is a poetic reference to the fish's upturned gaze. Other stargazers are called *Uranoscopus*, or "looking at the heavens." The Mediterranean stargazer, known to the Ancient Greeks and Romans and the first stargazer to be given a scientific name, carries the specific name *scaber*, meaning rough, scurvy or untidy. Though this is an apt description of its appearance, its flesh is relished in many countries. Divers tend to steer clear of it, for although its electric shock is not dangerous, its shoulder spines can inflict painful wounds that may become infected.

# STARLING

TAKEN AS A FAMILY, the starlings are highly successful but none can compete with the success story of the common starling, *Sturnus vulgaris*. It is 8¼–8⅗ inches (21–22 cm) long with a stout body, a short tail and broad, triangular wings. Its plumage changes markedly throughout the year. Young starlings are dull brown, and in their first autumn they molt to a spotted plumage with a brown head. At the same time the adults have many white spots on a dark ground with a green and purple iridescence, the spots of the female being bolder than those of the male. By the spring the spots are gone as a result of the abrasion of the ends of the feathers, and all adults are then blackish, with iridescence. The bill is quite long and horn-colored in winter and yellowish in spring and summer. The changes are confusing, but when its plumage is at its best the common starling is a handsome bird. So also are many of its relatives.

## A colorful family

There are 113 species of Old World birds in the starling family, Sturnidae. These include the African oxpeckers or tickbirds and the mynahs or mynas, both of which are discussed elsewhere. All are medium-sized perching birds, usually with loud, harsh voices and variable plumages.

The rose-colored starling, *Sturnus roseus*, of eastern Europe and Central Asia, has a bright pink body, a dark-crested head and black wings and tail. The glossy starlings of the South Pacific are greenish black as adults but are heavily streaked with white when young. Some species, such as the African wattled starling, *Creatophora cinerea*, have further ornamentations. This species molts its head feathers in the breeding season and grows long wattles.

## Living clouds

Starlings tend to live in flocks, and this is most marked in the common starling. During the day each flock spreads out for feeding, but in the late afternoon all the common starlings in the neighborhood begin to come together for roosting. The pattern of behavior then varies. The birds typically begin to gather in small groups of 12 to 20 on bushes or in trees. Each group later joins a

nearby group, and the process of forming larger and larger groups continues until a flock thousands strong is formed, which flies around and around, spreading out and coming together like a huge smoke cloud in the sky.

## Orderly roosting formations

Sometimes a flock of several hundreds will fly in formation, turning, wheeling and changing course with almost military precision. At other times the birds will make directly for the roost in small groups of two or three or up to a dozen. There are times when common starlings gather in trees in a noisy chorus, which is traditionally known as a murmuration. Then suddenly, as if cut by a knife, the chorus stops and a few seconds later all the birds take to wing and fly off. As they fly away, the last five or six birds return to the tree and start singing again. Other starlings fly in from all directions until 40 to 50 have assembled and a chorus builds up again.

*An adult common starling arriving with a billful of food for its chicks. Starlings usually nest in holes in trees, often taking over the nests of martins and woodpeckers.*

*In winter, common starlings are covered with white spots. Like many other species in their family, they are sociable birds that form large flocks.*

Then suddenly it ceases again, as if cut off with a knife, and a few seconds later they all fly off in the same direction as the previous group, again a few birds detaching themselves from the rear and returning to the tree. The whole ceremony is reenacted perhaps as many as six times before all the starlings in the surrounding area have assembled and flown to the main roost.

## Nuisance in towns

In rural areas starlings roost in a clump of trees, fouling the ground beneath with their droppings, or in a church tower in a village. This was the pattern of their roosting everywhere until the 1890s, when they started to roost in towns. The first record in London was in 1894. Every night in many towns and cities, except in the breeding season, a vast flock assembles, the individual birds perching on ledges and windowsills for the night. On one occasion starlings settled in such numbers on the big hand of Big Ben that they stopped the clock. Starlings roosting on buildings have now become a regular feature, especially in larger towns. Many methods of driving them away have been tried to save the buildings from being fouled by their droppings, but they have all met with indifferent success.

## Colonizing America

The common starling, which is native to most of Europe, North Africa, the Middle East and western and Central Asia, has been taken to Australia, North America and southern Africa. The species' most spectacular spread has been in North America. Several unsuccessful attempts had been made to introduce the birds into the United States and Canada. Then 60 were released in a New York park in 1890, and 40 more in 1891.

## COMMON STARLING

| | |
|---|---|
| CLASS | **Aves** |
| ORDER | **Passeriformes** |
| FAMILY | **Sturnidae** |
| GENUS AND SPECIES | ***Sturnus vulgaris*** |

WEIGHT
**2–3⅙ oz. (57–90 g)**

LENGTH
**Head to tail: 8¼–8⅔ in. (21–22 cm); wingspan: 14½–16½ in. (37–42 cm)**

DISTINCTIVE FEATURES
**Fairly long, pointed bill and sloping forehead, giving very slender appearance to front of head. Breeding adult: blackish plumage with strong iridescent greenish and purple gloss. Nonbreeding adult: covered with white spots. Juvenile: brown plumage; pale throat.**

DIET
**Wide variety of invertebrates (especially insect larvae); also seeds, grain and berries**

BREEDING
**Age at first breeding: 1 year; breeding season: eggs laid March–July (Northern Hemisphere); number of eggs: usually 4 to 6; incubation period: 12–13 days; fledging period: about 21 days; breeding interval: 1 or 2 broods per year**

LIFE SPAN
**Up to 15 years**

HABITAT
**Most habitats except dense forest and high mountains, including towns and cities**

DISTRIBUTION
**Native to Eurasia, Middle East and North Africa; introduced to North America, southern Africa and Australia**

STATUS
**Very common**

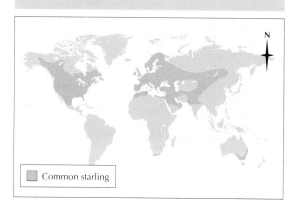

Common starling

By 1948 they had spread across the United States and reached the Pacific coast. They have now spread to Canada and Mexico.

## Mainly beneficial to agriculture

In the 1920s a panel was set up in the United States to assess the effect of common starlings on agriculture. Its report showed that the starlings are beneficial because of what they eat, and this slightly outweighs their nuisance value in other ways, such as fouling the ground under their roosts and raiding soft fruits. For most of the year starlings probe the grassland and plowed fields with their bills for insect grubs, especially for the troublesome wireworms. They also feed among cattle, taking insects disturbed by the animals' hooves. At other times common starlings take soft fruits, and in autumn and winter they feed on berries, gorging themselves as long as a particular crop of berries lasts.

## Hole nesters

Most starlings nest in holes in trees. The common starling also nests in holes in buildings and in eaves, and it often will drive other birds out of suitable nesting cavities. Breeding begins in March or April, the male building the nest of leaves and dry grass and the female lining it. The eggs, usually four to six in a clutch, are very pale blue, sometimes with small red spots. They hatch in 12–13 days, the female incubating them at night and the two sharing the incubation by day.

The fledglings are fed by both parents for about 3 weeks. More than most birds, starlings tend to lay occasional eggs on the ground. Polygamy has been recorded, with one male mating with three females and helping each of them with the incubation and feeding of the fledglings.

## Success against enemies

Much of the success of common starlings in building up large populations and spreading over new areas is due to their adaptability in using nesting sites, their wide diet and their own pugnacity—they will drive other birds from feeding tables. It also is due to the close-knit family life: starlings more than most birds seem to control and marshal their young, especially in times of danger. Attacks by birds of prey on flocks of adults are largely thwarted by the starlings flying in tightly packed box formations as soon as the enemy is spotted.

## Sounds that deceive

Most species of starlings are very vocal, but besides their native calls many are proficient mimics of the calls of other birds or of mechanical sounds. When a gull or shorebird is heard calling and neither can be seen, it is likely to be a common starling on the rooftops imitating their calls. Once, when the fountain in the garden of an English house had been turned off, the tinkle of falling water could still be heard; eventually it was traced to a starling.

*First introduced in New York in the late 19th century, the common starling has since spread across most of the United States and Canada.*

# STEAMER DUCK

*Like all steamer ducks, the Falkland Island steamer duck has white wing patches. It is found only on the Falklands, where it is also known as the logger duck.*

THE FOUR SPECIES OF steamer ducks are large birds that live in southern South America and on the nearby Falkland Islands. The body is very heavy, weighing up to 14 pounds (6,350 g) in the largest species, the feet are large and the bill is very broad. The wings are short, and three species of steamer ducks are practically unable to fly.

The Magellanic flightless steamer duck, *Tachyeres pteneres*, is the largest of the four species. Males weigh an average of 11⁷⁄₁₀ pounds (5,310 g) and the females weigh slightly less at 9½ pounds (4,328 g), but otherwise the sexes are quite similar. The plumage is mottled gray. The head and neck are gray mottled with white, the crown is bluish gray and there is a brownish tinge on the throat. In common with the females of the other two flightless steamer ducks, females of this species have a darker head than males.

The Magellanic steamer duck lives on the coasts of South America from Concepcion in Chile to Cape San Diego in Argentina, including the islands around Tierra del Fuego. The Falkland Island flightless steamer duck, *T. brachypterus*, is smaller than the Magellanic species, and the sexes have similar plumage. Both male and female have darker bodies than the Magellanic steamer duck, with a yellowish collar around the neck. Females have a dark head with a white stripe running back from the eye; males have a paler gray head than females.

The white-headed steamer duck, *T. leucocephalus*, was described as recently as 1981. It is found only on the coast of Chubut, a region in southern Argentina. The contrast between the white head of the male and its grayish body is the strongest of any steamer ducks. The female has a broad pale crescent running from behind the eye to a pale collar. This contrasts with a chocolate brown head.

The flying steamer duck, *T. patachonicus*, is the smallest of the four species, with the males weighing an average of 6⅔ pounds (3,030 g). It is found both in the Falkland Islands and in southern South America, from Valdivia in Chile and Puerto Deseado in Argentina south to Tierra del Fuego. The ranges of the three flightless steamer ducks do not cross over, although the flying steamer duck's range overlaps with all of them. The flying steamer duck has longer wings and a longer tail than the flightless species and its legs are more slender. Its plumage is almost brown over most of the body; the male can be distinguished from the female by a whitish head. All steamer ducks have white patches on their wings that can be seen when the wings are folded.

## Steaming across the water

Steamer ducks are so called because of their habit of steaming over the water, rushing across the surface propelled by their wings and legs and throwing up sheets of spray. The Magellanic steamer duck has been recorded as steaming at 8 miles per hour (13 km/h) over short distances when chased by a boat. The flying steamer duck can fly well but prefers to steam if disturbed. Many large males are incapable of flight because they are so heavy.

Under certain conditions, all three of the so-called flightless species can fly short distances, but they do not rise far above the surface. The flightless steamer ducks are confined to coasts and are rarely seen in fresh water except to drink and bathe. The flying steamer duck, however,

# FLIGHTLESS STEAMER DUCK

CLASS **Aves**

ORDER **Anseriformes**

FAMILY **Anatidae**

GENUS AND SPECIES **_Tachyeres brachypterus_**

ALTERNATIVE NAMES
**Falkland Island flightless steamer duck;
logger duck**

WEIGHT
**Male: average 9½ lb. (4,334 g);
female: average 6¾ lb. (3,383 g)**

LENGTH
**Head to tail: 24–29 in. (61–74 cm)**

DISTINCTIVE FEATURES
**Large size; mostly gray, but with white belly
and white wing patch; yellow bill; yellow
legs. Male: pale gray head. Female:
chocolate-brown head; whitish eye-ring;
white stripe running behind eye.**

DIET
**Marine mollusks; some crustaceans**

BREEDING
**Age at first breeding: 1 year; breeding
season: eggs laid mostly September–
December; number of eggs: 5 to 10;
incubation period: 34 days; fledging period:
about 84 days; breeding interval: 1 year**

LIFE SPAN
**Not known**

HABITAT
**Shallow coastal waters, rarely far from land;
often visits freshwater pools near shore in
winter; nests close to shore**

DISTRIBUTION
**Widespread throughout Falkland Islands**

STATUS
**Common**

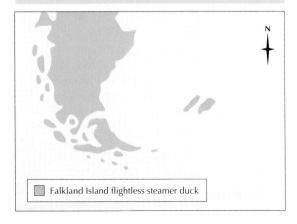

Falkland Island flightless steamer duck

regularly inhabits lakes and rivers. Coastal steamer ducks are rarely seen farther than a few hundred feet from land, and the Magellanic steamer duck is not found on the parts of the South American coast that have very large tides.

In the Falkland Islands, where steamer ducks have few if any predators, they are remarkably tame. In Port Stanley, the only town in the Falkland Islands, flightless steamer ducks live on the beach just below the main road.

## Feeding and breeding

Steamer ducks feed on aquatic animals, which they catch by diving in shallow water. The diet is mainly mollusks but also includes crustaceans, although the flying steamer duck, with its smaller bill, eats fewer thick-shelled mollusks than the other species.

In their breeding behavior steamer ducks show several similarities to shelducks (discussed elsewhere in this encyclopedia), to which they may be related. They mate for life and the male vigorously defends the territory, not only against other steamer ducks but also against other birds, including other kinds of ducks, penguins and geese. Battles between male steamer ducks may be very violent. They swim toward each other, sinking lower and lower into the water, so when they meet they are almost submerged. They fight by grabbing each other's heads and beating with their wings, which bear yellow knobs.

The 5 to 10 eggs are laid in a nest near the shore and are incubated by the female while the male keeps watch, either from the shore or from the water. Shortly after the chicks have hatched, the female leads them to the water. They stay near the female but the male is always nearby and comes to the female's assistance if the brood is molested by gulls or skuas. Some chicks, however, fall prey to these birds.

*Falkland Island flightless steamer ducks have 5 to 10 young. The female (above) is discernable from the male by the white stripe running back from her eye.*

# STEENBOK

The steenbok can be distinguished from other antelopes such as the bush duiker by its large round ears and, in males, the upright, polished horns.

THE STEENBOK IS A DELICATE antelope with large eyes and ears, which makes it very appealing. With its relative the grysbok, it is closely related to the oribi and the suni. The steenbok and grysbok are 24 inches (60 cm) or less in height and have a coarse, rough coat. The males have simple, spiky horns. Whereas the oribi and suni have ribbed horns set at an angle to the vertical, the steenbok and grysbok have smooth upright horns. Like the suni, they lack the knee tufts and the bare patch under the ear of the oribi, the tail is short and not bushy and the underparts are white. The male has a small gland in the groin, and both sexes have a rounded gland 2 millimeters in diameter and 3 millimeters deep just in front of the eyes.

The steenbok is a uniform reddish or grayish fawn and has no lateral hooves, just the small subsidiary hooves that many antelopes have, above and behind the main one. The two species of grysbok are smaller and have white-speckled coats and longer ears. The ordinary grysbok has small lateral hooves and is deep rufous, whereas Sharpe's grysbok is tawny rufous with no lateral hooves. In Sharpe's grysbok, as opposed to the other two, the face gland is surrounded by very short hairs; in the other two it is surrounded by a large area of naked skin.

The steenbok has a discontinuous distribution: the southern race is found from southernmost Africa, except for the southeastern coast, north as far as the Zambezi and southern Angola. The northern race, which is paler with white rings around the eyes, is found in western Tanzania and as far north as southwestern Kenya. The grysbok is found in the coastal region of the southwestern Cape, as far east as 28° E. Sharpe's grysbok is found from Natal and Transvaal north into Tanzania, reaching as far north as 2° 30' S in the west of Tanzania.

## Hermits of the bush

Steenbok and grysbok both live in scrub and bush country wherever there is enough undergrowth to hide them. Though the two species overlap a little, the grysbok tends to replace the steenbok in woodland, the latter favoring savanna, especially in the south of its range. Both species colonize areas of secondary growth, for example where elephants have foraged or agricultural land has been abandoned.

By day these antelope lie up in the grass or in old aardvark holes, only emerging at dawn and dusk to feed and move about. They are usually solitary. Generally the largest group seen is three, probably a mated pair with its most recent young. Each individual (or pair, when two do come together) occupies a territory. Whether this is actually defended is not known, but the occupant does not appear to wander outside it once the territory is established. The territory is marked by dung heaps, which have the owner's scent on them and are scattered at points around the territory.

The steenbok browses close to the ground on the shoots of native trees such as *Acacia* and *Combretum*, also scratching up roots and tubers with its hooves and nibbling on seedpods. Grass becomes more important after rains or burning, constituting two-thirds of the diet in some areas. The grysbok also eats fruits and seedpods, switching to tougher foods in the dry season.

## Ceremonial courtship

The female steenbok comes into season every three months for about four days at a time. When she does, the male becomes very aggressive,

# STEENBOK AND GRYSBOK

CLASS **Mammalia**

ORDER **Artiodactyla**

FAMILY **Bovidae**

GENUS AND SPECIES **Steenbok, *Raphicerus campestris*; grysbok, *R. melanotis*; Sharpe's grysbok, *R. sharpei***

ALTERNATIVE NAMES
**Steenbuck, steinbuck, steinbok (*Raphicerus campestris*); Cape grysbok (*R. melanotis*)**

WEIGHT
**15½–35 lb. (7–16 kg)**

LENGTH
**Head and body: 24–38 in. (60–95 cm); shoulder height: 18–24 in. (45–60 cm); tail: 1½–3¼ in. (4–8 cm)**

DISTINCTIVE FEATURES
**Small antelope with russet coat and short, pointed horns; white belly and tail; steenbok more slender than other two species**

DIET
**Mainly leaves; also grasses, herbs, roots, fruits and seeds**

BREEDING
**Age at first breeding: 6–20 months; breeding season: throughout year; gestation period: 170–210 days; number of young: usually 1; breeding interval: 1 year**

LIFE SPAN
**Up to about 10 years**

HABITAT
**Scrubland, savanna, thickets and riverbeds**

DISTRIBUTION
**Steenbok: discontinuous range in southern and East Africa; grysbok: southeastern Africa**

STATUS
**Steenbok: locally common; Sharpe's grysbok and grysbok: conservation-dependent**

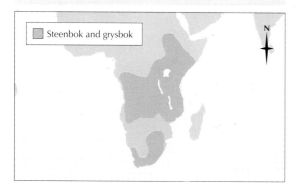

Steenbok and grysbok

using his horns in territorial and rivalry disputes. Two paired adults rub their faces together, nibbling at each other and exchanging secretions from their face glands, like duikers. The courtship ritual also includes a *Laufschlag*, or leg-beat. Gestation is 210 days; the female lies down to give birth, and in the wild it seems that birth takes place in a disused aardvark burrow, which is usually the only available cover on the savanna. There is a birth peak in November and December, at least for the southern steenbok; there is usually a single young.

Young Sharpe's grysbok were recorded to be 10 inches (25 cm) high at three weeks and to weigh 2–3 pounds (0.8–1.2 kg); this is somewhat more than half the adult height and a quarter of the adult weight—a large youngster, in fact, for such a small antelope. Nibbling on grass within two weeks of birth, the young is weaned off its mother's milk by three months. It is sexually mature from as early as six months.

## Escape tactics

Leopards, cheetahs, jackals and hawks are likely to be the most dangerous predators on these little antelope. The grysbok's usual response to danger is to lie hidden in the grass with outstretched neck. If it is seen, it scuttles swiftly and fluidly away with its head straight out in front, sometimes zigzagging, but seldom running far before ducking again into the undergrowth or an abandoned aardvark burrow. The steenbok also lies low, flattening its huge ears and following an intruder's every move. When flushed out, it runs with more of a gallop and holds its head higher. When cornered, a steenbok often puts up a fight, striking out with its hooves.

*Leaves and shoots of savanna trees are on the steenbok's menu, which also includes anything from seedpods, berries and fruits to buried roots and tubers.*

# STEPPE LEMMING

THE STEPPE LEMMING, *Lagurus lagurus*, is commonly used by scientists in laboratory experiments. It is native to the Ukraine and Mongolian steppes. A second, similar species, known as the sagebrush vole, *Lemmiscus curtatus*, combines the characteristics of lemmings and voles and is found in southwestern Canada and the northwestern United States. However, it is not a true steppe lemming.

The steppe lemming is a stocky rodent 3–5 inches (8–13 cm) in length, with a tail ¼–¾ inch (7–20 mm) long. The sagebrush vole's tail is slightly longer, at ⅗–1⅕ inches (16–30 mm). The steppe lemming has the blunt muzzle and the small ears and eyes of a vole, and its face is covered with numerous very long whiskers. Its long, soft upper fur is cinnamon to gray in color with white underparts. There is a black stripe along the back. The legs are short, the toes have stout claws and the soles of the feet are hairy.

Unlike the related rodents, lemmings and voles, steppe lemmings live mainly in semi-deserts and dry steppes where grass is sparse. However, they may also move into pastures and arable land, where they are regarded as pests. Steppe lemmings are mainly nocturnal, only

*Both the steppe lemming (pictured below) and the sagebrush vole are regarded as pests in agricultural habitats.*

occasionally being seen by day, and they are active year-round. Sagebrush voles are active both by day and night.

Steppe lemmings live in loose colonies. They make short burrows, each of which consists of several tunnels with a number of entrances and several nesting chambers. Most desert rodents eat mainly seeds but steppe lemmings feed chiefly on green plants, tubers and roots. As long as there is plenty of green food, they do not need to drink. Laboratory animals live best on grass and hay with thin twigs of willow, seeds, grain and a limited amount of root vegetables, such as carrots and beets. If they are given too rich a diet, they can become overweight and their breeding rate drops as a result.

## Unintentional haymakers

Steppe lemmings do not move far from their burrows. During the summer, when grass is most plentiful, they leave some grass lying on the ground wherever they feed. Some individuals, an average of about 1 in 10, take this hay litter into their burrows to store as food. The rest of the colony use this store in winter when supplies are short, taking it as and when needed.

## STEPPE LEMMINGS

| | |
|---|---|
| CLASS | **Mammalia** |
| ORDER | **Rodentia** |
| FAMILY | **Muridae** |

GENUS AND SPECIES  **Steppe lemming, *Lagurus lagurus*; sagebrush vole, *Lemmiscus curtatus***

**WEIGHT**
⅗–1½ oz. (17–42 g)

**LENGTH**
Head and body: 3–5 in. (8–13 cm).
Tail: (*L. lagurus*) ¼–¾ in. (7–20 mm);
(*L. curtatus*) ⅗–1⅕ in. (16–30 mm).

**DISTINCTIVE FEATURES**
Cinnamon gray upperparts with black dorsal stripe; white underparts

**DIET**
Shoots, grasses, leaves and some seeds

**BREEDING**
Age at first breeding: 2–3 years; breeding season: year-round, peaks in April–October; gestation period: (*L. lagurus*) 20 days, (*L. curtatus*) 25 days; number of young: 5 to 9; breeding interval: 5 to 8 litters per year

**LIFE SPAN**
Up to 3 years

**HABITAT**
Grasslands, prairies and arid steppes

**DISTRIBUTION**
*L. lagurus*: Ukraine and Mongolian steppes.
*L. curtatus*: southwestern Canada and northwestern U.S.

**STATUS**
Common

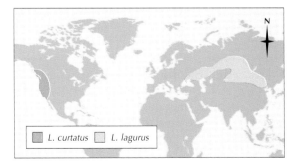

L. curtatus ☐ L. lagurus

Wasteful feeding and food hoarding are common among rodents. However, a few animals, such as the pikas (discussed elsewhere in this encyclopedia), can be said to make hay. The steppe lemmings may perhaps be in the process of acquiring the habit.

### Large infant turnover

Steppe lemmings make their nests out of plant fibers. The sagebrush vole breeds year-round. The true steppe lemming has about five litters during the summer and some in winter. Its gestation period is 20 days, while that of the sagebrush vole is 25 days. There are usually five to nine young in a litter. Unusually for rodents, nursing female steppe lemmings are very tolerant of each other and often live together when they have litters. In the laboratory a pair may have 10 to 12 litters a year. The babies weigh 1.4 grams at birth. This is doubled in 5 days and doubled again in 12 days, by which time the eyes have opened and the babies have begun to feed themselves. They are fully weaned 5 to 6 days later. The females are sexually mature at 60 days, the males at 60–75 days. The life span is 2–2½ years, with a maximum of 3 years.

### Valued laboratory animals

Steppe lemmings were first bred in the laboratory 30 years ago in the former Soviet Union. In the early 1960s, colonies were established in Germany and England for research on diseases such as tularemia and poliomyelitis. Scientists subsequently realized that the species has only a small number of chromosomes and that the individual chromosomes are easily distinguished.

The lemmings are regarded by scientists as an asset in the study of genetics, specially the genetics of cancer. They are small, easy to feed and maintain, do not hibernate, are easy to tamed and are docile in nature. Moreover, steppe lemmings breed rapidly and grow quickly, so there is a rapid turnover in terms of individuals. In their rapid succession of generations, steppe lemmings have something of the advantages of vinegar flies, *Drosophila*, with regard to genetic research, and they are smaller than guinea pigs.

*Although steppe lemmings are mainly nocturnal, sagebrush voles are also active during the day, which makes them vulnerable to predators, such as coyotes.*

# STICK INSECT

*Stick insects are skillful copyists, closely mimicking the twigs and stalks among which they live. Pictured is a stick insect of the genus Bacteriidae.*

STICK INSECTS ARE TODAY more common pets than probably any other insect. They are sluggish and live among the foliage of trees and bushes or in low-growing herbage, relying for protection on their ability to mimic their surroundings. They are long and very slender, and have long, thin legs. Stick insects usually have smooth bodies, although some species are spiny.

The larger types of stick insects resemble twigs and may be green or brown; the small species and the young of the larger ones are usually green and resemble the midribs of leaves or the stems and blades of grass. Some stick insects are very large. The Asian species, *Palophus titan*, for example, is the longest insect alive today, sometimes exceeding 12 inches (30 cm) in length. Some stick insects have wings, but many are wingless, a condition that enhances their resemblance to twigs.

Stick insects, along with the leaf insects, comprise an order, the Phasmida, that scientists once included in the Orthoptera together with the grasshoppers, mantids, cockroaches and others. However, this group has subsequently been divided into several separate orders.

About 2,000 species of phasmids are known, the majority being found in the Asian Tropics. One species, *Bacillus rossii*, is native to Europe, ranging as far north as central France. Two types of stick insects from New Zealand have become established in the extreme southwest of the British Isles, where the climate is generally warmer: the prickly stick insect, *Acanthoxyla prasina*, in Devon and on Tresco in the Scilly Isles, and the smooth stick insect, *Clitarchus hookeri*, also on Tresco and on an island off County Kerry, Ireland.

The laboratory stick insect, *Carausius morosus*, is an Asian species often kept in schools, laboratories or as a pet and can adjust its color to match its background. It is a very easy insect to keep and breed and can be fed on leaves of privet, ivy or lilac, but it cannot survive outdoors through the cold winter in northern Europe and must be kept inside.

## Flash coloration

Most stick insects feed and move about only at night. By day they remain motionless and often appear to be feigning death. In fact, they pass into a hypnotic or cataleptic state during the day. When they are in this condition, the limbs can be moved into any position and will stay there, rather as if the joints were made of wax. Some of the winged species are active by day. In many of these the hind wings, which are the only ones developed for flying, are brightly colored but are entirely concealed when the insect is at rest. If the insect is disturbed, the wings are suddenly unfolded and the resultant flash of bright color is confusing to a searching predator. Then, when the wings are closed again, the bright color suddenly disappears, so the precise position at which the insect has alighted is effectively concealed. This well-known protective device is known as called flash coloration.

All stick insects are plant eaters and they occasionally become numerous enough to defoliate areas of woodland. In Australia there are two species that occur in swampy areas but also feed on agricultural crops, sometimes causing serious damage.

## Numerous eggs

All stick insects lay rather large, hard-shelled eggs that look very much like seeds. In some cases they closely resemble the actual seeds of the plant on which the insect feeds. The eggs are dropped by the females at random and the tapping sound of falling eggs is often heard from the cages of captive stick insects. One North American species of stick insects, *Diapheromera femorata*, is sometimes so numerous that the sound of thousands of its eggs falling on the forest floor may be as loud as that of rainfall.

The female stick insects usually lay several hundred eggs, a few being laid each day, and they can take a long time to hatch. Those of the laboratory stick insect hatch in 4–6 months at ordinary room temperatures. However, this can be speeded up to 2 months by extra warmth or retarded to 8 months by cold conditions, such as

---

| STICK INSECTS | |
| --- | --- |
| PHYLUM | **Arthropoda** |
| CLASS | **Insecta** |
| ORDER | **Phasmida** |
| FAMILY | **Phyllidae, Bacteriidae, Phasmatidae** |
| GENUS AND SPECIES | **About 2,000 species, including giant prickly stick insect, *Extatosoma tiaratum*; smooth stick insect, *Clitarchus hookeri*; and laboratory stick insect, *Carausius morosus*** |

**LENGTH**
**Usually 3–4 in. (8–10 cm); some species over 12 in. (30 cm) long**

**DISTINCTIVE FEATURES**
**Long, slender, twiglike insects; long legs; some species winged, others wingless; young resemble adults**

**DIET**
**Plant material**

**BREEDING**
**Many species have no known male sex, males rare in other species; parthenogenetic reproduction (reproduction without sexual intercourse); hundreds of eggs laid; hatching period: variable, from a few weeks up to 1 year, depending on conditions; larvae undergo several molts before reaching adulthood**

**LIFE SPAN**
**Up to several years**

**HABITAT**
**Usually found in vegetation**

**DISTRIBUTION**
**Most species in tropical regions; some in Europe and New Zealand**

**STATUS**
**Common**

---

*Giant prickly stick insect,* Extatosoma tiaratum, *instar. When hatched, young stick insects are miniature adults, and undergo several molts before reaching adult size.*

*Although most stick insects have long, cylindrical bodies resembling stalks and sticks, some have more flattened bodies that look like small twigs.*

an unheated room in winter. The eggs of the Madagascar stick insect, *Sipyloidea sipylus*, will hatch in as little as 1 month if kept at 75–80° F (24–27° C). At lower temperatures, however, the eggs may lie dormant for up to 12 months. The young look very much like the adults in every respect except size and, in the case of the winged species, in lacking wings, which develop gradually during growth.

Most, possibly even all, stick insects reproduce by parthenogenesis, that is, the females lay fertile eggs without mating. In these species the males usually are rare; in cultures of the laboratory stick insect, for example, males number about one to every 4,000 females. Of the two New Zealand species already mentioned, the male of the prickly stick insect is unknown and possibly does not exist. In New Zealand, males of the smooth stick insect are almost as common as females, but no males have been found in the small British colonies of the same species, and the eggs develop without fertilization.

## Color changes

The laboratory stick insect occurs in various color forms, ranging from green to shades of brown. The color is determined by green, brown, orange-red and yellow granules in the cells of the surface layer of the skin. Pure green individuals cannot change color, but other stick insects regularly change, becoming darker at night and paler by day. The change is brought about by movement of the pigment granules within the cells. Brown pigments may move to the surface and spread out, making the stick insect dark in tone. Alternatively, these pigments may contract into lumps and move to the inner part of the cell so that the insect becomes pale. The orange-red granules can also move about in this way, but not the green and yellow ones.

The stick insect's alternation of colors becomes established by exposure to normal day and night. However, once established, the color alternation continues as a rhythm governed by the time cycle of 24 hours. A stick insect conditioned to normal light change and then kept in permanent darkness will continue for several weeks to change color every 24 hours, just as it did before. If it is kept in the dark by day and exposed to artificial light at night, a reversed rhythm will develop in response to these conditions. This rhythm also persists for some time when the insect is kept continually in darkness, with no light at all.

# STICKLEBACK

CIENTISTS USED sticklebacks in some of the earliest modern studies of animal behavior, and today they are used in water pollution tests. All sticklebacks have a long body, large head and strong jaws. They range in size from 2½–7 inches (6–18 cm), most being only 3–4 inches (7.5–10 cm) long. The largest is the sea stickleback, *Spinachia spinachia*. They are usually greenish to black in color on the back and silver on the belly, sometimes with dark bars on the sides. They have two dorsal fins, the first of which is made up of well-spaced spines. The anal fin is similar to the second dorsal and lies opposite it. Each pelvic fin is one long spike and the pectoral fins are large. Most sticklebacks have a series of bony plates along each flank; the number varying with the species and according to temperature and salinity.

There are 10 species and several subspecies in the North Temperate Zone of the Northern Hemisphere, and two of them range across Europe, Asia and North America. Sticklebacks are tolerant of salty water, at least two species being found in the sea as well as in fresh water, and two are wholly marine.

## At home in river or sea

The three-spined stickleback or tiddler, *Gasterosteus aculeatus aculeatus*, is the most widespread stickleback and is the one described here. It occurs throughout the Northern Hemisphere and lives in all fresh waters except fast-flowing mountain streams. It also is found in estuaries and along the coasts, and it has been caught 2–3 miles (3.2–4.8 km) out at sea. It is not often found in stagnant or weed-choked waters, where the nine-spined stickleback, *Pungitius pungitius pungitius*, can live. The sea or 15-spined stickleback is wholly marine. The 2½-inch (6.3-cm) brook stickleback, *Culaea inconstans*, is found in the fresh waters of the United States and Canada, and the four-spined stickleback, *Apeltes quadracus*, is common along the eastern seaboard, from Virginia as far north as Nova Scotia.

## Armored or not

The variation in the bony plates or scutes along the flanks of the *G. aculeatus* species complex has led scientists to name four types. The trachurus type swims up rivers from the sea to spawn. This type has a complete row of scutes from head to tail and is found in the north of the range, in salty waters and usually in half-grown individuals only. In the same areas live the semiarmata type, with scutes halfway along the body, and

the leiurus type, which has scutes only on the abdominal side. The gymnurus type inhabits fresh waters in England and France. These have no scutes and are found in the south of the range.

## Mixed carnivorous diet

Sticklebacks feed on almost any small invertebrate, the size of the prey depending on the age of the fish. It includes small crustaceans such as water fleas and freshwater shrimps, worms, small mollusks and their larvae, aquatic insects and their larvae and sometimes fish eggs. Corresponding marine invertebrates are taken by sticklebacks living in salt water, which grow more quickly and to a slightly greater maximum size than those living in fresh water. Despite their spines, sticklebacks are themselves preyed upon, specially by kingfishers and grebes. They also tend to be infected with tapeworm; in some lakes all sticklebacks may be affected in this way.

## Nest-building fish

As the breeding season approaches, the male becomes more brightly colored, with red on the front part of the underside. He is known as a red throat at this time. The male takes over a territory and drives out intruding sticklebacks. In the center of the territory he builds a nest of small pieces of plants glued together with a sticky secretion from his kidneys. Sea sticklebacks use

*There are several stages in a stickleback's courtship display. Here, the pale, red-throated male urges the female to enter the nest and lay her eggs.*

*The male stickleback guards the eggs and cares for the young when they hatch. He also ventilates the nest by fanning water through it.*

pieces of smaller seaweeds for their nests. The nest is lodged among the stems of water plants or among seaweeds in the sea. When ready, the male entices one or more females to lay her eggs in it. As each female lays and leaves, the male enters the nest and sheds his milt to fertilize the eggs, which are just under 2 millimeters in diameter. The eggs hatch in 5–12 days, according to the temperature. During this time the male aerates them by fanning water through the nest. The baby sticklebacks, ⅙ inch (4 mm) long when hatched, are guarded by the male until they are ready to leave the nest. They grow to a length of 1–2 inches (2.5–5 cm) in the first year. The life span in the wild is about 3 years.

## Study in courtship

The three-spined stickleback first gained prominence after the Dutch animal behaviorist Nikolaas Tinbergen (1907–1988) studied its courtship. The study provides a simple illustration of the use by animals of sign stimuli. A male stickleback guarding his territory attacks another male because it has a red throat. Even a wooden stickleback model held in a stickleback's territory will be attacked, provided it has a red throat marking. A female entering the territory and ready to lay, turns her abdomen swollen with ripe eggs toward the male. On seeing this, he swims excitedly in what is called a zigzag dance. He responds in the same way to a wooden model having the same shape. Having danced to her, the male turns and swims toward the nest, pointing his head at the entrance. The female follows and enters and the male butts her in the flank with his snout and trembles, which makes her respond by laying. After she has left the nest

### STICKLEBACKS

| | |
|---|---|
| CLASS | **Osteichthyes** |
| ORDER | **Gasterosteiformes** |
| FAMILY | **Gasterosteidae** |
| GENUS | **5 genera** |
| SPECIES | **10 species, including three-spined stickleback, *Gasterosteus aculeatus aculeatus* (detailed below)** |

LENGTH
**About 4 in. (10 cm)**

DISTINCTIVE FEATURES
**Long body; large head; dark brown-green upperparts; silvery sides; 3 long spines on back. Breeding male: bright red throat.**

DIET
**Worms, crustaceans and aquatic insects; also small fish and own eggs and young**

BREEDING
**Engages in nest-building and courtship dance; breeding season: spring–early summer; number of eggs: 50 to 100; hatching period: 5–12 days**

LIFE SPAN
**Up to about 3 years**

HABITAT
**Vegetated fresh waters, usually over mud or sand; in sea, confined to coastal waters**

DISTRIBUTION
**North Atlantic; much of Europe; North Africa; North Pacific, from Korea north to Bering Sea**

STATUS
**Locally common**

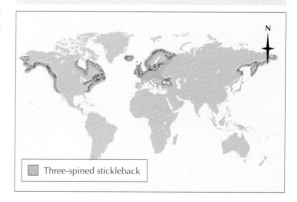

Three-spined stickleback

the male enters and fertilizes the eggs. The spawning is the result of a series of orderly stereotyped actions, each step being triggered by a definite signal or sign stimulus, such as the red throat, swollen abdomen and zigzag dance. However, the spawning routine may vary at times.

# STICK-NEST RAT

STICK-NEST RATS RIVAL THE pack rats (discussed elsewhere) of America in the large nests of sticks they build. Early colonists in Australia called them native rats or rabbit-rats because their relatively large ears and blunt noses gave them a resemblance to small rabbits sitting hunched up. The two species of stick-nest rats, *Leporillus apicalis* and *L. conditor*, vary from 5½–8 inches (14–20 cm) in length, and the long hairy tail, slightly tufted at the tip, may be up to 10 inches (25 cm) long in *L. apicalis*. The fur is thick and downy, the upperparts varying from light yellowish and dull brown to pale grayish brown, with gray or white underparts.

Stick-nest rats live in south-central Australia, and *L. conditor* also lives on Franklin Island off the coast of South Australia. *L. conditor* is distinguished by having shorter ears and not such thick fur, which is colored dark amber brown on the back. *L. apicalis* has a slighter build, a paler grayish brown back, white fur underneath and a white-tipped tail. The mainland species were once abundant but are probably now confined to a few areas remote from human habitation; they may even be extinct. The island species, however, is flourishing and has a much better chance of survival than those on the mainland.

## Stronghold against predators

The stick-nest rats are unusual for their habit of building nests of sticks for shelter and breeding. Some of these nests are communal and house large colonies. The nests vary in size and structure, according to the local conditions. Those of *L. conditor* are usually constructed around a bush, and the sticks are strongly interwoven among the stems and branches of the bush. The nests are built up to a height of about 3 feet (0.9 m) and may be up to 4 feet (1.2 m) in diameter, sometimes larger. They are constructed with great care and form a stronghold against the dingo and carnivorous marsupials as well as against high winds. In the center of the larger nests are several soft grass nests with numerous entrances and passages leading to them. In areas where the bushes are too small or weak to be used for supports, the nests take the form of loose heaps of sticks placed over rabbit warrens, the tunnels of which provide the animals with an easy means of escape. The stick-nest rats work stones into these unsupported nests and other stones are placed on top to weigh down the nests, anchoring them against high winds.

On Franklin Island, *L. conditor* sometimes builds enormous nests of sticks and debris on the top of the cliffs, housing extra-large colonies of rats. One such nest was built on the abandoned nest of a sea eagle as a foundation. Larger nests are sometimes located over penguin burrows, and it has been observed that on the approach of danger, penguins, shearwaters, bandicoots and even black tiger snakes may bolt into the burrows and tunnels with the rats. On the shore or the flatter parts of the island the nests are small, housing usually only one pair of rats and are made of dried herbage or seaweed. In small nests there is only one chamber and one or two entrance tunnels; on the shore the nest may be no more than dried seaweed tucked between large stones.

Apart from reports dating back to 1864, very little is known of the habits of *L. apicalis*, but it seems that it does not consistently build nests. Sometimes it will shelter in hollow trees or the deserted nests of *L. conditor*. It is gregarious, like *L. conditor*, and also nocturnal.

## Vegetarian diet

Stick-nest rats are believed to have a totally vegetarian diet. *L. conditor* on Franklin Island feeds mainly on the leaves of a plant, *Tetragona*.

*The greater stick-nest rat, Leporillus conditor, now lives only on Franklin Island, off Deduna, southwest of South Australia.*

However, zoologists suspect that they may sometimes eat the eggs and young of birds. For example, there is a record of the nest of a striped brown hawk being built on top of a nest of *L. conditor*. Although the hawk did not molest the rats, there were signs that the rats raided the nest when it was left unguarded.

## Small litters

The breeding habits of stick-nest rats remain a subject of scientific debate. The young are born in the soft grass nests in the center of the stick nests. The female has four teats so probably, unlike most rodents, she has a fairly small litter of perhaps only one or two young. The young of *L. conditor* are carried about by the mother, hanging from her teats, which they grasp firmly in their mouths, a habit that originally gave rise to the mistaken idea that the rat was a pouchless marsupial. This habit, however, is seen in a number of small rodents in different parts of the world.

## A variety of predators

Dingoes find it hard to penetrate the stick-nest shelters, and the chief predator of stick-nest rats seems to be the masked owl, *Tyto novaehollandiae*. It preys specially on *L. conditor* on the desolate Nullarbor Plain, where the owl lives in the numerous limestone caves. In the past, Aborigines hunted stick-nest rats for their flesh. In settled parts of New South Wales *L. apicalis* is thought to have been exterminated by introduced foxes and domestic cats. Snakes and introduced predators also prey on stick-nest rats.

## Early observations

The stick-nests of *L. conditor* were first observed on a surveying expedition into the interior of eastern Australia in 1838. When the explorers first saw the numerous piles of sticks on the plains of the Murray and Lower Darling in New South Wales, they thought they were piles of brushwood put there by Aborigines for their signal fires. They had no reason to suspect rats because at the time this habit was not known to be characteristic of some rodents. Closer examination of these carefully constructed nests and the fact that the kangaroo dogs scratched and barked at them encouraged the explorers to break open a nest, which they did with difficulty. Inside they found the soft nests containing small animals that they described as resembling a small rabbit, apart from the tail. Unfortunately, they did not keep careful watch on any animal in the nest and reported that the nests were made by the white-footed rabbit-rat, *Conilurus albipes*, a rodent then common in New South Wales and Victoria, and a relative of the stick-nest rats. It was not until 1844, during an expedition into

### STICK-NEST RATS

| | |
|---|---|
| CLASS | **Mammalia** |
| ORDER | **Rodentia** |
| FAMILY | **Muridae** |
| GENUS AND SPECIES | **Leporillus conditor;** **L. apicalis** |

WEIGHT
**1–1⅖ oz. (25–40 g)**

LENGTH
**Head and body: 5½–8 in. (14–20 cm); tail: up to 10 in. (25 cm)**

DISTINCTIVE FEATURES
**Yellow, brown or gray upperparts; gray (*L. conditor*) or white (*L. apicalis*) underparts**

DIET
**Plant material, seed leaf and shoots**

BREEDING
**Age at first breeding: 8 months; breeding season: year round; gestation period: 44 days; number of young: 1 or 2; breeding interval: opportunistic and variable**

LIFE SPAN
**Up to 4 years**

HABITAT
**Dry woodland, scrub and heaths**

DISTRIBUTION
**South-central Australia and Franklin Island**

STATUS
**L. conditor: endangered; about 5,000 survive on Franklin Island. L. apicalis: probably recently extinct.**

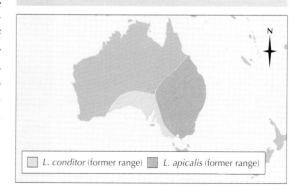

☐ *L. conditor* (former range)  ▨ *L. apicalis* (former range)

central Australia, that it was realized that the animal was a distinct species, and it was given the specific name of conditor, meaning a maker or contriver. A specimen of the rat was brought back and a full account of the animal and its unusual nest-building habits, illustrated by drawings, was provided.

# STIFFTAIL

STIFFTAILS ARE A GROUP OF ducks that have their tails made up of stiff feathers. They are relatively small ducks, 13–18 inches (33–45 cm) long, according to species. Stifftails have a characteristic short, thickset neck, which can be inflated by an air sac under the esophagus. The bill is broad, the wings are short and the webbed feet are large. The plumage is often finely spotted and barred, with the underparts being white mottled with brown, but there are no metallic sheens, as are found in many other ducks. Stifftails occur in most parts of the world, but the majority of them are poorly known.

## Eight species

Among the better known stifftails are the ruddy duck, *Oxyura jamaicensis*, of the Americas, and the white-headed duck, *O. leucocephala*, of Eurasia.

The ruddy duck ranges from central Canada south to Tierra del Fuego, being found over most of North and Central America and the Caribbean, but it is restricted to the region of the Andes Mountains in South America. In its breeding plumage the male is bright chestnut, with a black crown and distinctive white cheeks. It also has a brilliant blue bill, which is gray in the female. The South American population can be divided into two subspecies, the Peruvian ruddy duck, which lacks white cheeks, and the Colombian, in which cheeks are mottled. The very rare white-headed duck, which breeds in a much reduced range in parts of southern Europe, Central Asia and North Africa, has a similar appearance to the ruddy duck, but the male has a pure white head and is less chestnut above, with a much heavier blue bill.

*A male ruddy duck in breeding plumage, Washington. Stifftails are named for their distinctive fan-shaped tails, which are often held vertically.*

*To attract a mate in the breeding season, the male ruddy duck slaps its bill against its chest to make a drumming sound, thus creating bubbles in the water.*

There are six other species of ducks in the stifftail group. The rare masked duck, *Oxyura dominica*, is a tropical species and a rare visitor to southern Texas and the Gulf coastline as far east as Florida. The breeding male has a black or dark brown face, but in nonbreeding plumage it looks very similar to a female ruddy duck. The maccoa duck (*O. maccoa*) of southern and eastern Africa and the blue-billed or spiny-tailed duck (*O. australis*) of Australia are like the ruddy duck but lack the white cheeks. The remaining species are the white-backed duck (*Thalassornis leuconotus*) of Africa, the musk duck (*Biziura lobata*) of Australia, which has a strange lobe under the bill, and the black-headed duck (*Heteronetta atricapilla*) of South America.

## Grebelike ducks

Stifftails are usually found on fresh water, preferring lakes, ponds and marshes where there is plenty of food. They are good divers, resembling grebes in their behavior and associate with them rather than with other ducks. If alarmed, stifftails can slowly submerge until only the head is showing and then disappear without a ripple. Their awkward progress on land also recalls grebes and loons. They cannot walk properly because their legs are placed well back, so they shuffle along on their bellies.

The dependence of stifftails on water is further shown by the difficulty of introducing them into captive waterfowl collections. Stifftails suffer on long journeys from not being able to swim. Although they have difficulty in takeoff, they fly well and some species migrate. Their flight is labored because their wings are small in relation to their heavy bodies.

## Dive to reach water plants

The food of stifftails is aquatic plants and animals. The white-headed duck eats mainly leaves and seeds, as well as fly larvae, water

## RUDDY DUCK

| | |
|---|---|
| CLASS | **Aves** |
| ORDER | **Anseriformes** |
| FAMILY | **Anatidae** |
| GENUS AND SPECIES | ***Oxyura jamaicensis*** |

WEIGHT
**Average 17⅔–21 oz. (500–590 g)**

LENGTH
**Head to tail: 13½–15 in. (34–38 cm); wingspan: 23–24 in. (58–60 cm)**

DISTINCTIVE FEATURES
**Chunky duck with large head and thickset neck; long, stiff, fan-shaped tail, often held vertically. Breeding male: bright blue bill; black crown and tail; white cheeks; rest of plumage red brown. Nonbreeding male and female: gray bill; dull gray-brown plumage with paler cheeks.**

DIET
**Mainly pond weed and other aquatic plants; also midge larvae**

BREEDING
**Age at first breeding: 1–2 years; breeding season: May–July (Canada and U.S.); number of eggs: 5 to 15; incubation period: about 25 days; fledging period: about 60 days; breeding interval: 1 year**

LIFE SPAN
**Up to 12 years**

HABITAT
**Lakes, ponds and marshes; in winter also shallow coastal bays and salt marshes**

DISTRIBUTION
**Native range: central Canada south to Tierra del Fuego, including the Caribbean; introduced range: parts of western Europe**

STATUS
**Common**

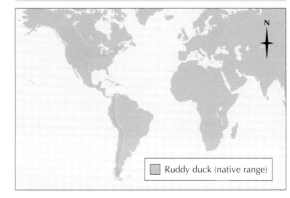

Ruddy duck (native range)

snails and crustaceans. The ruddy duck eats mainly pond weed. However, the musk duck is most unusual in that it has a predominantly carnivorous diet, including crayfish (known as yabbies in its native Australia) and crabs.

## Helpful males

Because stifftails nest in thick reed beds in inaccessible places, their breeding behavior is poorly known. In general, their nests are elaborate and well hidden. The eggs are very large, sometimes as big as those of ducks three times their size. The male helps to care for his family, although he does not help with incubation.

The courtship behavior of male stifftails with their bright plumage is quite spectacular. They inflate their throat sacs so they look like pouter pigeons and tilt their tails over their backs, repeatedly pressing the bill into the inflated breast and clucking or squeaking. The male Argentinian ruddy duck has been described as bringing its bill down on its inflated chest with such force that it produces a drumming sound audible up to about 50 yards (46 m) away.

The nest is built among sedges, reeds or bulrushes, which are arched over the nest for better concealment. The white-headed duck sometimes builds floating nests, but more often the nest is firmly woven into standing plants. The white-headed and blue-billed ducks sometimes take over the abandoned nests of coots, grebes and other ducks.

The ruddy duck lays up to 15 eggs, which together may weigh 3 pounds (1.4 kg), the equivalent of three times the weight of the duck that laid them. The usual clutch is smaller but up to 20 have been found in one nest. The chicks hatch in about 3½ weeks, and shortly after they are escorted to the water by both parents. They can dive for food almost immediately.

## Strange ducks

The Australian musk duck is peculiar because of the leathery pouch or flap under the bill, which is larger in males than in females, and because of the musk gland, which secretes an unpleasant odor in the breeding season. Not surprisingly, musk ducks are not good to eat. Another name for the musk duck is mold goose.

Another species with unusual habits is the black-headed duck. It is not very closely related to the other stifftails. It probably is a brood parasite, because its nest has never been found, but eggs that are very likely to be those of the black-headed duck have been found in the nests of other birds, even in those of caracaras.

*The white-headed duck (male, below) is now very rare in much of its range. One of the main threats it faces is interbreeding with introduced ruddy ducks. Interbreeding gradually reduces the species' genetic purity.*

# STILT

*In flight the long legs of stilts trail behind the tail. Pictured is a black-winged stilt in the Gambia.*

THE STILTS BELONG TO the same family of shorebirds as the four species of avocets (genus *Recurvirostra*) and the single species of ibisbill (*Ibidorhyncha struthersii*). While the bills of avocets are turned up at the tip and that of the ibisbill is turned down, those of the stilts are very thin and straight. Stilts have proportionately longer legs than any other shorebirds. Their head and body length is about 13–15 inches (33–38 cm), according to species.

## Global range

The black-winged stilt, *Himantopus himantopus*, is black and white and has a very large range, with a variety of common names to describe its different geographic subspecies, or races. It breeds in southern Europe, France, Africa, Asia and Australasia. There are only a few breeding records from Britain. In Australia and New Zealand the species is known as the pied or white-headed stilt. Other alternative names include longshanks and stiltbird. The species is migratory in the north of its range.

## Other species

The species found in the Americas, the black-necked stilt, *H. mexicanus*, has a continuous black band from the crown to the back and a distinctive white spot above the eye. It breeds from southern Canada and Oregon south to northern South America, including the Galapagos Islands and the Caribbean region. In the fall most of the populations that breed in the United States move south for the winter.

The third species is the banded stilt, *Cladorhynchus leucocephalus*, which is found only in Australia. It is black and white with a chestnut band across the breast. While black-winged and black-necked stilts have pink or reddish legs, those of the banded stilt are pinkish orange or orange yellow. The critically endangered black stilt, *Himantopus novaezelandiae*, is now restricted to a single valley in South Island, New Zealand.

The ibisbill lives in the high mountains of Central Asia at altitudes of 5,600–14,400 feet (1,700–4,400 m). It frequents shingle banks and small islets, usually in fast-moving streams.

## BLACK-NECKED STILT

| | |
|---|---|
| CLASS | **Aves** |
| ORDER | **Charadriiformes** |
| FAMILY | **Recurvirostridae** |
| GENUS AND SPECIES | ***Himantopus mexicanus*** |

WEIGHT
**Average 6 oz. (170 g)**

LENGTH
**Head to tail: about 14 in. (36 cm);
wingspan: 21–22 in. (53–56 cm)**

DISTINCTIVE FEATURES
**Long, thin, black bill with needlelike point;
red irises; extremely long, pink or reddish
legs that extend well beyond tail in flight.
Adult male: black crown, hind neck and
upperparts; white forehead, spot above eye,
sides of neck and underparts. Adult female
and juvenile: brownish black upperparts.**

DIET
**Aquatic insects, worms, water snails,
tadpoles and tiny fish; also seeds**

BREEDING
**Age at first breeding: 2 years; breeding
season: April–June (Canada and U.S.);
number of eggs: usually 4; incubation
period: 24–27 days; fledging period: about
28 days; breeding interval: 1 year**

LIFE SPAN
**Up to 15 years**

HABITAT
**Shallow margins of freshwater lakes, rivers
and marshes; brackish (slightly salty)
lagoons; salt marshes; flooded fields**

DISTRIBUTION
**Breeds from southern Canada south to
South America, including Galapagos Islands
and Caribbean region; winters south of U.S.**

STATUS
**Locally common**

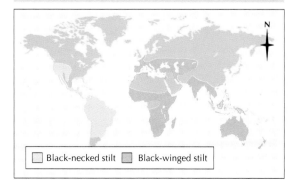

☐ Black-necked stilt ☐ Black-winged stilt

## Prefer weedy shallows

Black-winged and black-necked stilts live in
pairs or small flocks along the margins of lakes,
marshes and slow-flowing rivers, and on flooded
agricultural land. They prefer shallow water
where there is plenty of submerged waterweed
and low-growing plants and avoid water overgrown with tall reeds. These two species also live
on salt marshes and brackish (slightly salty)
lagoons. In the United States, the largest counts
of black-necked stilts occur in the Great Salt Lake
area, Utah; Stillwater National Park, Nevada;
and Brazoria National Wildlife Refuge, Texas.

The banded stilt usually lives on temporary
salt lakes, often far from the coast in the arid interior of Australia. Occasionally it is seen in
estuaries, but it appears to shun fresh water.

## Standing tall

Stilts walk with a slow, graceful gait, picking up
their slender legs and placing them delicately in
long strides. In flight, stilts have rapid wingbeats
and they carry their heads held in. On long
flights they trail their legs behind the tail, making
an unmistakable silhouette. When they are man-
euvering, stilts use their long legs as rudders.

*A black-winged stilt
searches for food in the
shallows of a lake in
Greece. Stilts walk
slowly with a graceful
gait, taking long,
leisurely strides.*

## Picking and probing

Most of the stilts' food is picked up from water plants or from the surface of the water, but they also probe in the mud as they wade in shallow water up to their knees and occasionally up to their bellies. This simple feeding behavior is reflected in the unspecialized straight bill, which contrasts with the upturned bill of avocets, used for sweeping the mud.

Stilts eat a wide variety of food: mainly insects such as water beetles and fly larvae taken from the water, as well as worms, water snails, tadpoles and tiny fish. Banded stilts depend on the brine shrimps and other small crustaceans that live in salt lakes.

## Nesting colonies

Stilts nest in colonies near their feeding grounds, and sometimes the nests are built in shallow water. Nests may be no more than scantily lined hollows, especially if in a tussock of grass, or they may be substantial structures of plants and mud. The clutch of eggs, which usually numbers three or four, is incubated by both parents. The chicks hatch out in 24–27 days and leave the nest site soon afterward. Their parents defend them by flying around intruders and performing very noisy distraction displays.

The nests of the banded stilt were not discovered by Western scientists until 1930, when a colony was discovered at Lake Grace in the state of Western Australia. Since then many other colonies have been found, some with tens of thousands of stilts, but nesting does not occur regularly at any one place. The species is dependent on the rainfall in an area, as this controls the availability of food and the suitability of the nesting sites. For instance, several nesting attempts at Lake Grace since 1930 have failed through flooding.

## Critically endangered

At one time the black stilt of New Zealand was thought to be a black form of the black-winged or pied stilt. However, it is now recognized as a true species in its own right. In the 19th century black stilts bred on shingle riverbanks throughout South Island and in much of North Island. In the 1930s they were still common in parts of South Island, but their numbers then suffered a dramatic decline. The black stilt is now confined to a river valley in the Mackenzie Basin area of South Island. In 1993 this population was estimated at about 60 birds.

Habitat loss is one of the main causes of the black stilt's decline. Its nesting areas have been destroyed by drainage and by hydroelectric and tree-planting projects. It also suffers from heavy predation because its exposed nest sites are vulnerable to introduced cats and ferrets. Hybridization with local black-winged stilts has also started to reduce the genetic purity of the remaining population.

*Stilts (black-necked stilt, below) have 8–10-inch (20–25-cm) legs, proportionately longer than those of any other shorebirds.*

# STINGRAY

STINGRAYS CAN INFLICT serious and poisonous wounds. Related to skates, they have a flattened body with winglike pectoral fins and a whiplike tail bearing a long poison spine. The disclike body may have a rounded leading edge or it may be drawn out slightly into a pointed snout. The pectoral fins and the pelvics are also rounded. The surface of the body is smooth with few or no denticles (toothlike, pointed projections). The tail is slender and at least as long as the rest of the body. The spiracles (breathing holes) are larger than the eyes. There is no dorsal fin, and the most obvious feature is the spine set in the tail, about a third of the way along. The upper surface is usually olive brown, sometimes with white spots or darker marbling, while the undersurface is white to creamy white. Stingrays measure from 12 inches (30 cm) to 14½ feet (4.4 m) across the fins and weigh from 1½–750 pounds (0.7–340 kg).

There are five genera of stingrays comprising a total of about 64 species. Stingrays live in tropical and temperate seas, as far north in the summer as southern Scandinavia in Europe and equivalent latitudes elsewhere. Stingrays live in shallow seas, seldom descending deeper than 400 feet (120 m). Some stingray species enter estuaries and swim up rivers, sometimes for considerable distances.

## Rapidly acting poison

Stingrays, like skates, spend much of their time on the seabed, searching for prey or merely resting. They move by wavelike undulations passing along the two pectoral fins; the tail is not used for swimming.

When it is attacked, or even if it has simply been disturbed, the stingray lashes with its tail, from side to side in some species, or moving the tail up and over the body in others. Such actions bring the swordlike spine into play. The spine grows up to 15 inches (38 cm) in length in the largest of the rays, and has saw-toothed edges and grooves. The grooves are lined with a glistening white tissue, which probably contains the stingray's poison.

The stab from a stingray injects poison and also cuts and tears flesh. Many people who have stepped on a stingray lying in shallow water have required stitches in their feet. Even a tiny puncture from the spine of a stingray is enough to cause a human to faint. The effect of the poison is immediate, and inflammation spreads around the wound almost as soon as the spine has penetrated. Other immediate symptoms are sharp shooting pains and throbbing. The poison affects the heart, breathing and nerves, and it can be fatal, although the remedies are simple and effective, provided they are applied in time.

*The blue-spotted stingray,* Taeniura lymma, *is widespread in the tropical Indian and Pacific Oceans. The people of the South Pacific once used stingray spines for their spearheads.*

*Stingrays can grow a new spine to replace an old one. Pictured is the black-spotted stingray or giant reef ray,* Taeniura melanospilos.

## STINGRAYS

| | |
|---|---|
| CLASS | **Chondrichthyes** |
| ORDER | **Rajiformes** |
| FAMILY | **Dasyatidae** |
| GENUS | **5 genera** |
| SPECIES | **64 species, including southern stingray, *Dasyatis americana* (detailed below)** |

ALTERNATIVE NAMES
**Clam-cracker; stingaree**

WEIGHT
**Up to 213 lb. (97 kg)**

LENGTH
**Up to about 6½ ft. (2 m)**

DISTINCTIVE FEATURES
**Diamond-shaped disc shape, with straight leading edges; ridge fold on tail from spine to tip; olive-brown above; whitish below**

DIET
**Mainly bivalve mollusks and worms; also shrimps, crabs and small fish**

BREEDING
**Ovoviviparous (young born alive); number of young: 3 to 4**

LIFE SPAN
**Not known**

HABITAT
**Sandy bottoms, seagrass beds, lagoons, reefs**

DISTRIBUTION
**Coastal waters of western Atlantic**

STATUS
**Common, especially in Caribbean**

Southern stingray

At one time, washing the wound with iodine or permanganate of potash was recommended. Today, the most effective form of treatment is to clean the wound, immerse it in hot water for up to an hour and have an antitetanus injection.

## Rows of teeth

The mouth of the stingray is located on the undersurface of the head. The jaws are wide and both have rows of blunt teeth, arranged so that there are several rows of broad teeth in the middle and rows of smaller teeth on each side. A North American species, the southern stingray, *Dasyatis americana*, is also called the clam-cracker or stingaree. Like other stingrays, this species feeds mainly on mollusks and crustaceans, sometimes preying on fish.

## Born alive

All stingrays are ovoviviparous (the eggs are not laid but hatch in the oviduct, the young eventually being born alive). At first, the young feed on the yolk in a yolk sac hanging from their abdomens, the food passing directly into the digestive tube from the sac. Then, at a later stage, blood vessels grow out and around the yolk and food is taken into the blood. Later, tiny filaments grow out from the walls of the oviduct. Each filament has a network of tiny blood vessels and gives out a liquid food, which the embryo stingray takes through its mouth or spiracles. This is the equivalent of the placenta in mammals.

## Specialized toothlike armor

Rays and sharks do not have scales as bony fish do. Their skin is protected by dermal denticles, each of which is made up of a pulp cavity inside a layer of dentinelike material with a type of enamel on the outside. Denticles were once thought to be the equivalent of real teeth, but scientists have now realized this is not the case. Scientists have not yet established whether the spine of a stingray is a modified and greatly enlarged dermal denticle. However, like true teeth the spine may be replaced by a new one.

# STONE CURLEW

THE STONE CURLEWS OR thick-knees are strange-looking shorebirds, the nine species making up the family Burhinidae. The former name is due to their preference for dry pebbly ground, although they are not closely related to the true curlews, discussed elsewhere. The latter name is due to the birds' swollen tibiotarsal (knee) joints.

Stone curlews range in size from 14 to 21 inches (36–53 cm) long. The legs are fairly long, the feet are slightly webbed and the hind toe is missing. The bill usually is short and thick, yellow or green with a black tip, and the eyes are very large. The plumage is dull gray or buff with streaks of brown and black.

The Eurasian stone curlew, *Burhinus oedicnemus*, is 16–17⅓ inches (40–44 cm) long, and has sandy brown plumage with black streaks and white wing bars. The bill is short and straight. It lives in Europe, including southern England, from Poland and Germany south to the Mediterranean, in North Africa and in southern Asia. Other stone curlews include the water dikkop or water thick-knee (*B. vermiculatus*) of Africa, the

great thick-knee (*Esacus recurvirostris*) of India, China and Southeast Asia, and the beach thick-knee or beach stone curlew (*E. magnirostris*) of Australia. The beach thick-knee is 21 inches (53 cm) long and has a massive bill, which is slightly flattened and upturned. Other stone curlews or thick-knees are found in tropical and temperate regions, including Central and South America, but they are missing entirely from North America, the islands of the Pacific and New Zealand.

## Shorebirds far from water

Although they are shorebirds, most stone curlews are found well away from water, often in dry upland regions or in sandy country. The bush thick-knee, *Burhinus grallarius*, is found in the light wooded and open country of Australia; the double-striped thick-knee, *B. bistriatus*, is found in the savannas of Central America and South America; and the South American thick-knee, *B. superciliaris*, is found in the sandy deserts of Peru and Ecuador. As the lack of a hind toe suggests, these birds are strong runners, and

*A Eurasian stone curlew incubating its eggs. The nest is no more than a scrape in the ground.*

*The stone curlews or thick-knees live in sandy or rocky habitats such as dry grassland, steppe, sand banks and gravel plains.*

there is some indication that they are more closely related to bustards than curlews. When the bush thick-knee, for instance, is chased, it seeks safety by running, and if it does take to the air, it has to taxi to take off. Then it lands quickly and runs into cover.

## Nocturnal feeders

Stone curlews are nocturnal, as is suggested by their large eyes, the pupils of which contract considerably during the day. At night the stone curlews are very noisy, producing cries that are often mournful. During the day they are quiet and lie up in cover. When disturbed, they flatten themselves with their head and neck extended; so with their drab plumage they are hard to find.

Stone curlews feed on insects, such as beetles, grasshoppers and fly maggots, as well as on snails, slugs, worms and crustaceans. Small rodents, the chicks of game birds, amphibians and reptiles are also eaten.

## Prefab nests

At the start of the breeding season male Eurasian stone curlews display vigorously, running about with outstretched wings. No nest is made, and the two eggs are laid in a bare scrape. At the start of incubation the parent stone curlew quietly creeps away from the nest when disturbed, but later it sits tight. Both parents incubate the eggs, which hatch in 24–26 days. The chicks leave the scrape shortly after hatching.

The water dikkop is one of the so-called crocodile birds. It associates with crocodiles during the nesting season. Both animals breed on sandbanks when the rivers are low, often only a few feet from each other. The birds gain protection from nest predators by the crocodiles' proximity. The crocodiles are in return given warning of enemies by the water dikkop's alarm calls.

## EURASIAN STONE CURLEW

| | |
|---|---|
| CLASS | **Aves** |
| ORDER | **Charadriiformes** |
| FAMILY | **Burhinidae** |
| GENUS AND SPECIES | ***Burhinus oedicnemus*** |

**WEIGHT**
**Usually about 1 lb. (455 g)**

**LENGTH**
**Head to tail: 16–17⅓ in. (40–44 cm); wingspan: 30⅓–33½ in. (77–85 cm)**

**DISTINCTIVE FEATURES**
**Large head and eyes; short, thick, yellow bill with black tip; long yellow legs with swollen knee joints; slightly webbed feet; pale sandy buff upperparts with subtle markings; white underparts with dark streaks on breast**

**DIET**
**Mainly invertebrates; also some small vertebrates, such as rodents, bird nestlings, frogs and lizards**

**BREEDING**
**Age at first breeding: not known; breeding season: eggs laid February–June (south of range), April–July (north of range); number of eggs: usually 2; incubation period: 24–26 days; fledging period: 36–42 days; breeding interval: usually 1 year**

**LIFE SPAN**
**Probably up to 16 years**

**HABITAT**
**Dry grassland, steppe, sandy shores of lakes and rivers, sand dunes and semidesert**

**DISTRIBUTION**
**Southern half of Europe, east through Middle East and Central Asia to India and Southeast Asia, south to North Africa**

**STATUS**
**Uncommon; serious population decline in agricultural regions, especially in Europe**

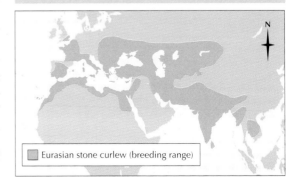

Eurasian stone curlew (breeding range)

# STONEFISH

THERE ARE 28 SPECIES of stonefish, and one of these, the estuarine stonefish, *Synanceia horrida*, is the most poisonous fish known. A stonefish is 6–15¾ inches (15–40 cm) long, has a heavy head that is broad and flat and a body that tapers rapidly from behind the head to the small tail fin. The mouth is wide and has a fairly large gape. The pectoral fins are large and winglike. The dorsal fin, which runs along the midline of the back, is armed with 13 stout spines. There are three more spines on the anal fin and one on each of the pelvic fins. The scaleless skin is covered with many irregular warts and a layer of slime. The color of the fish is similar to that of mud, seaweed or stone. One stonefish species even resembles a piece of rock covered with small algae.

There are two species of the world's most venomous fish: *S. horrida* and *S. verrucosa*. They range from the Red Sea to East Africa and across the Indian Ocean south to the northern coasts of Western Australia and Queensland, Australia.

## Defenses on all sides
Stonefish live in shallow seas, especially where the bottom is coral rock or tidal mudflats. They lie completely still even when approached by a human, and if a person places a foot a few inches from them, the stonefish's only reaction is to erect their spines.

The stonefish is virtually invisible against its background, and its victims probably never see it before it attacks. Each of the spines has two poison sacs near its tip. Pressure on this tip makes a sheath covering it slide back, leaving the point of the spine bare and exposing the grooves down which the poison flows. Although usually it is the spines on the back that do most damage, if the fish is rolled onto its side it can still defend itself using the anal and pelvic fins.

Fishers in the Indian Ocean handle stonefish with great care, specially as the fish can stay alive 10 hours after they have been taken out of water. Even dead specimens lying on the beach are still able to inflict a poisonous wound.

## Waiting for food
Stonefish wait for their food to come to them. Any passing animal that is not too large to be swallowed, mainly fish and crustaceans, is rapidly snapped up. The stonefish never use their poison spines for catching food, but only as a means of self-defense.

The stonefish's breeding habits remain a subject of scientific debate. Scientists know rather more about the fish's predators, although it may seem surprising to learn that such a poisonous fish does have predators. Certain bottom-feeding sharks and rays, with crushing teeth used for eating crabs and hard-shelled mollusks, occasionally take stonefish. The stonefish's habit of staying very still puts it at a distinct disadvantage with regard to some of its predators. In tropical seas there are large sea snails called conchs (discussed elsewhere), which are both aggressive and carnivorous, and stonefish, especially young ones, are eaten by them.

## Thirteen deadly spines
Reports of the effect of stonefish spines on humans differ. Some suggest that people have trodden on or handled stonefish and been either unaffected or little the worse for the experience. At the other extreme, there are reports of fatalities as a result of an encounter with the fish. It seems that a person can be very slightly pricked in the finger and, provided the sheath is not broken or the wound is only shallow, no poison will be injected. It also seems that once the spines have been touched and their poison ejected they are harmless, suggesting that the poison sacs, the

*The stonefish's mottled and textured skin gives it the appearance of an algae-covered rock lying on the bottom. This camouflage feature has given rise to its common name.*

## STONEFISH

| | |
|---|---|
| CLASS | **Osteichthyes** |
| ORDER | **Scorpaeniformes** |
| FAMILY | **Synanceiidae** |
| GENUS | **8 genera** |
| SPECIES | **28 species, including stonefish, *Synanceia verrucosa* (detailed below)** |

LENGTH
**Up to 15¾ in. (40 cm)**

DISTINCTIVE FEATURES
**Large head; wide-gaped mouth; 13 spines and 2 poison grooves on dorsal fin; fawn-brown upperparts, mottled and reticulated, brown or red spots; paler underparts**

DIET
**Fish and crustaceans**

BREEDING
**Not known; possibly planktonic eggs**

LIFE SPAN
**Not known**

HABITAT
**Sandy or rubbly areas of reef flats and shallow lagoons, small pools during low tide**

DISTRIBUTION
**Indian Ocean and South Pacific, from Red Sea and East Africa to French Polynesia, north to Ryukyu and Ogasawara Islands, south to Queensland, Australia**

STATUS
**Locally common**

*Perhaps unsurprisingly, considering its reputation as the world's most venomous fish, the estuarine stonefish has not been studied in any great detail.*

sheath or both cannot be renewed. In contrast, there are authentic cases on record of immediate, extremely painful symptoms, followed by death. These victims suffered excruciating pain, which caused them to scream with agony, and then collapsed into delirium. They died after about six hours. However, if the wounds are not fatal such severe pain may last up to eight hours and then slowly diminish. In some instances the legs swell excessively, there may be large blisters and the skin may slough.

Antidotes, which must be applied quickly, include a weak solution of hydrochloric acid or formalin and permanganate of potash. At the Serum Laboratories in Melbourne, Victoria, Australia, an antivenin has been produced.

### Aboriginal folklore

Any suspicion that reports of such grievous symptoms as a result of stepping on a stonefish are exaggerated must be set aside in view of a ritual performed by certain Australian Aborigines. It takes the form of a charade, which has been enacted since the time of the Bronze Age in Europe, from around 4000–3000 B.C.E. to 1000 B.C.E., that is to say, back in the Aborigines' dreamtime, an ancient period of their history.

In the charade, a dancer imitates a man paddling in the tidal pools looking for fish. He takes short steps, looking to the left and to the right. Then he takes big steps and suddenly lifts one foot, grabs it with a hand, screams and limps away. He sits down and then lies down, writhing and screaming, while a shaman dances around him uttering incantations. Finally, the shaman throws up his hands in despair and the patient

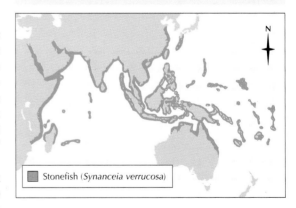

Stonefish (*Synanceia verrucosa*)

wails a death song. Perhaps the most notable feature of this theatrical display is that, throughout, the dancer carries a clay model of a fish with 13 splinters of wood stuck into its back to represent spines, a totem that symbolically represents a stonefish.

# STONEFLY

STONEFLIES BELONG TO the Plecoptera, one of the most ancient insect orders; fossil stone flies are known from the Permian period, about 250 million years ago. Plecopterans have aquatic larvae, and the adults, too, are generally found near water. There are around 2,000 species in the order.

Adult stoneflies have two pairs of membranous wings, long, threadlike antennae and a pair of segmented appendages, called cerci, at the hind end. The word cercus comes from the Greek, meaning tail, and in some insects the cerci carry organs of hearing. Stoneflies are weak fliers, and when at rest the large hind wings are folded like a fan and partially covered by the narrow forewings, one of which overlaps the other.

Stoneflies can be large; the North American *Pteronarcys californica* is 3½ inches (8.5 cm) in length. *Perla* and *Nemoura* are common European genera. Dark brown and dull green are the most usual colors, but some Australian species have rich red hind wings tinged with purple streaks. One species is jet black with orange markings. The larvae, or nymphs, are basically similar to the adults in structure, except that they lack wings and they have tufts of filamentous gills along each side of the abdomen for extracting oxygen from water. Their legs are often fringed with hairs for swimming.

## Living up to their name

Adult stoneflies are usually found resting on or under stones or on tree trunks near streams, preferring temperatures no higher than about 75° F (25° C). They are secretive and inactive, their muted colors making them nearly impossible to spot against the background. They tend to run, rather than fly, when disturbed. They have weakly developed biting mouthparts and are said to feed on algae, although the adults of many species may not eat at all but only drink.

The larvae also spend much of their time under stones on the riverbed. Most eat plant food, such as pollen or small algae grazed off the surface of stones, but larger species also eat the aquatic larvae of other insects. They especially eat dragonfly nymphs, pouncing on them as soon as their antennae or cerci detect the prey.

*A stonefly nymph (larva) perches on a rock in a stream. In all species the larvae are aquatic, using feathery gills to extract oxygen from the water.*

*A stonefly of the genus Perloida, photographed near Prague in the Czech Republic. Adult stoneflies typically spend the day resting on waterside plants.*

## STONEFLIES

| | |
|---|---|
| PHYLUM | **Arthropoda** |
| CLASS | **Insecta** |
| ORDER | **Plecoptera** |
| SUBORDER | **Archiperlaria, Filipalpia and Setipalpia** |
| FAMILY | **Nemouridae, Perlidae and others** |
| GENUS AND SPECIES | **At least 2,000 species** |

LENGTH
**Usually ⅕–⅔ in. (5–15 mm), but some species much larger**

DISTINCTIVE FEATURES
**Adult: medium-sized insect; flattened body; 2 pairs of membranous wings, of which hind pair are the larger; 2 long, threadlike cerci (tail-like filaments); 2 long antennae. Larva: broadly similar to adult but wingless.**

DIET
**Adult: pollen and algae. Larva: most eat plant matter, but some prey on aquatic insects.**

BREEDING
**Undergo partial metamorphosis; eggs laid on water; larva undergoes several molts**

LIFE SPAN
**Adult: 2–3 days to 2–3 weeks; life cycle usually complete in 1 year but can be longer**

HABITAT
**Adult: among waterside stones or vegetation. Larva: clean, unpolluted streams, particularly those with stony beds.**

DISTRIBUTION
**Worldwide apart from polar regions**

STATUS
**Very common**

## Aerial egg-laying

The females lay their eggs in a variety of ways. In some species the female swims on the surface, dropping her eggs as she does so. The female of one large species is a strong swimmer and leaves a V-shaped wake behind her. Other females fly over water, dipping the tip of the abdomen into the water every now and then. There is a third group in which females fly over water, alighting from time to time just to lay eggs and then flying away, or falling on the surface with their wings uplifted, releasing their eggs as soon as the tip of the abdomen touches the water.

## Long larval life

The eggs are dropped into the water in thin, membranous packets. The whole preadult life is passed in running water, which must be pure and well oxygenated because the larvae depend entirely on their feathery gills for respiration. This helps explain why stoneflies favor cool waters, which are rich in dissolved oxygen.

The aquatic preadult life lasts at least a year, and sometimes a period of up to four years is passed in the water. The wings develop gradually, and there are many molts, or ecdyses; up to 33 have been recorded. The larvae remain active at extreme low temperatures, down to almost freezing in some of the Arctic species.

In order to avoid the warmth of spring, the larva usually makes its final molt early in the year, even when ice still fringes its home waters. The molt tends to occur at night, and males usually emerge before females. When ready, the larva crawls from the water onto a stone or a tree, anchors itself with its hooked feet, and then splits down the back to release the winged adult, which eventually pulls itself free.

## Profit and loss

Stonefly larvae, also known by anglers as creepers, are of great importance as a source of food for trout and other fish that live in the clean, swift streams that they inhabit. Some anglers maintain, however, that the creepers are a mixed blessing because the larger individuals feed on the larvae that sport fish, such as trout, also eat.

Stonefly larvae are also useful as indicators of pollution. Their presence shows that a stream is almost or completely free from pollution; on the other hand, their absence from apparently suitable waters is a warning that some degree of pollution is probably present.

# STORK

STORKS WITH THEIR long necks, bills and legs, resemble their relatives, the herons and ibises. The plumage usually is black and white. The stork family Ciconiidae contains 17 species in six genera, including the five species of so-called typical storks in the genus *Ciconia*. The other genera are the wood storks (*Mycteria*, four species), the openbill storks (*Anastomus*, two species), the saddlebill and black-necked storks (*Ephippiorhynchus*, two species), the jabiru stork (*Jabiru*, one species) and the marabou and adjutant storks (*Leptoptilos*, three species). The marabou, *L. crumeniferus*, is discussed elsewhere.

The hamerkop, *Scopus umbretta*, and shoebill stork, *Balaeniceps rex*, are often known as storks but are classed in separate families from the true storks. Both of these African species are treated separately in this encyclopedia.

## A varied and widespread family

One of the most familiar storks is the white stork, *Ciconia ciconia*, which often appears in European legend. It stands nearly 3 feet (90 cm) high and has pure white plumage with black flight feathers and a bright red bill and legs. It breeds from the Netherlands and Denmark east through western Russia, Greece and Turkey to Iran, as well as in Spain, Portugal and northwestern Africa. Its range decreased in the 20th century, and it no longer nests in southern Sweden or Switzerland. Closely related to the white stork is the black stork, *C. nigra*, which is mainly black with white underparts. It breeds in the Iberian Peninsula, eastern Europe and across temperate Asia to China. There also are isolated populations in southern Africa.

Other Old World storks include the 4-foot (1.2-m) high saddlebill, *Ephippiorhynchus senegalensis*, of tropical Africa. The two openbills are smaller; one lives in tropical Africa and Madagascar and the other in southern Asia. The black-necked stork, *E. asiaticus*, is found in Africa, Asia and Australia. In the Americas there are two species in the Tropics: the huge jabiru (*Jabiru mycteria*), with a black, naked head, and the maguari stork (*Ciconia maguari*).

*The white stork's nest is a huge platform of twigs that can be seen from afar. Like other storks, white storks pair for life and use the same nest year after year.*

*A white stork and black stork feed side by side in the shallows. Storks are much less dependent on water than herons and ibises, often feeding in fields or open grassland.*

The best-known wood stork, *Mycteria americana*, ranges from North Carolina south to Argentina. It has bare, black skin on the head. In the Old World there is the African wood stork, *M. ibis*, which has a red face, and two Asian species that are known as painted storks. The wood storks are sometimes called wood ibises, but this can lead to confusion with the true ibises.

## Huge nesting platforms

Many storks nest in tall trees but some nest on cliffs, and the white stork often nests on buildings or pylons. The black stork nests among the branches of tall trees in woodlands, but other storks prefer open country. Where there are plenty of nest sites, storks nest in colonies, using the same nests year after year. They are large platforms of sticks, perhaps 6 feet (1.8 m) across. When a pair meets at the nest, the birds perform a characteristic display, clattering their bills and bending their necks back until the head is touching the back.

There are usually three to five eggs in a clutch. They are incubated by both parents for just over 1 month. Because incubation starts after the first one or two eggs have been laid and subsequent eggs are laid at 2-day intervals, the chicks hatch out one after the other, and the oldest is quite strong by the time the last emerges. As a result, the younger chicks often die unless there is an abundance of food. At first the chicks are given food, which is regurgitated by the parents, but later whole animals are given straight to them. The young storks leave the nest when about 8–9 weeks old.

## Long migrations

Although storks often live near water, where they wade in shallow pools and marshes in search of food, they are much less dependent on

### WHITE STORK

| | |
|---|---|
| CLASS | **Aves** |
| ORDER | **Ciconiiformes** |
| FAMILY | **Ciconiidae** |
| GENUS AND SPECIES | *Ciconia ciconia* |

**WEIGHT**
**5–9⅔ lb. (2.3–4.4 kg)**

**LENGTH**
**Head to tail: 3¼–3¾ ft. (1–1.15 m); wingspan: 5–5⅖ ft. (1.55–1.65 m)**

**DISTINCTIVE FEATURES**
**Heronlike body; long, powerful red bill; very long red legs; huge, broad wings; pure white plumage except for black flight feathers and tail**

**DIET**
**Large insects such as grasshoppers and locusts, snails, frogs, small fish, lizards, mice and voles; occasionally bird nestlings**

**BREEDING**
**Age at first breeding: usually 4 years; breeding season: late March–May (Europe); number of eggs: usually 3 to 5; incubation period: 33–34 days; fledging period: 58–64 days; breeding interval: 1 year**

**LIFE SPAN**
**Up to 25 years**

**HABITAT**
**Open wetlands, marshes, fields, grassland, savanna and steppe**

**DISTRIBUTION**
**Summer: central and northern Europe east to Iran; Spain, Portugal and northwestern Africa; eastern China. Winter: southern Europe, sub–Saharan Africa, western Middle East, India and southeastern China.**

**STATUS**
**Locally common**

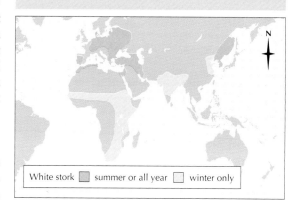

White stork ▢ summer or all year ▢ winter only

water than the herons and ibises. The maguari stork and several others feed in dry country such as savanna, and the marabou even visits urban areas to feed on refuse. Storks fly strongly, on broad, rounded, heronlike wings, with the neck extended and the legs trailing. Some storks perform long migrations, making use of thermals (rising bubbles of warm air) to soar to great heights with hardly a wingbeat.

The migration of the white stork is especially well known because the flocks of migrating birds are easily seen and large numbers have been banded by ornithologists. As with migrating birds of prey, long sea crossings are avoided, and European storks make their way to Africa across the Bosphorus and the Straits of Gibraltar, soaring high over the land and then gliding across the sea. The two streams pass through East Africa and West Africa, respectively, and it is known that the eastern stream finally ends up in eastern South Africa. A few white storks may stay in southern Africa to breed, and there is a resident population of black storks nesting on mountain faces in South Africa, presumably derived from European immigrants.

## Voracious appetites

Storks feed on a wide variety of animals caught in water or on land. In shallow water they catch small fish, pond snails, frogs and large insects. Large insects are also one of their main items of prey on land. Locusts and grasshoppers are particularly favored, and white storks will interrupt their migrations in East Africa when there is a plague of locusts. Larger land animals such as rodents, lizards and the young of ground-nesting birds are also eaten.

The two species of openbills are so named because when their bills are closed, although the two mandibles meet at the tips, there is a gap between them over most of their length. This is an adaptation for holding the large water snails that form a major part of their diet.

## Popular but declining

Because of the white stork's large size and its habit of nesting on houses, it has been possible to carry out detailed censuses over a long period. Unfortunately, these censuses show that the white stork, like many others, is declining due mainly to modern intensive farming methods.

*A flock of painted storks,* Mycteria leucocephala, *in India. Painted storks are relatives of the American wood ibis and like that species have a patch of bare skin on the face.*

# STORM PETREL

*In common with other storm petrels, Leach's storm petrel,* Oceanodroma leucorhoa, *is a pelagic (open sea) bird that comes onshore only to breed on offshore islands and rocky coasts.*

THE NAME PETREL IS thought to allude to St. Peter and refers to the ability of some petrels, including storm petrels, to patter across the water's surface for short distances. Storm petrels are so called because sailors once thought their appearance heralded a storm. Another nickname is Mother Carey's chickens, probably from Mater Cara, the Virgin Mary.

The storm petrels are the smallest members of the order Procellariiformes, which includes the albatrosses, shearwaters and fulmars (all discussed elsewhere). Storm petrels can be distinguished from other members of the family by the tubular nostrils on top of the bill, which are united to form a single opening. Compared with the large albatrosses and their 10–11-foot (3–3.3 m) wingspans, the storm petrels are midgets, being only 5½–7 inches (14–18 cm) long. They generally are dark, almost black, but some have white rumps or white underparts. Three species are gray.

The storm petrels fall into two groups: those living in the Southern Hemisphere, most of which have more rounded wings, very slender bills and very long legs; and those living mainly in the Northern Hemisphere, which have shorter legs and longer, pointed wings.

Among the many species of storm petrel, one of the best known is Wilson's storm petrel, *Oceanites oceanicus,* which breeds around the coasts and islands of Antarctica and migrates to the North Atlantic as far as Britain and Newfoundland. It is distinguished by the yellow webs between its toes. One of the best-studied storm petrels is the European storm petrel, *Hydrobates pelagicus,* one of the most restricted species, breeding in the North Atlantic and Mediterranean. Most storm petrels are found in the Pacific, particularly around the fertile areas of upwelling currents off Peru and Japan.

## Riding the storm

Storm petrels can sometimes be seen from the shore when their migration routes take them inshore, or when strong onshore winds blow them close to coasts. After storms, they may be blown inland and are sometimes seen over lakes far inland. Otherwise, they probably are best known as small, swallowlike birds that fly low over the wakes of ships, with an erratic batlike flight. Often, about half a dozen storm petrels follow a ship, presumably attracted by the small animals thrown to the surface by the churning of the ship's propeller. Apart from the Galapagos or wedge-rumped storm petrel, *O. tethys,* they are not visible at their breeding colonies by day. The colonies, on rocky island slopes, are visited only at night, when masses of storm petrels fly to and fro before landing near their burrows and exchanging twittering, crooning calls with their mates within. One storm petrel that does not nest on islands is Hornby's storm petrel, *O. hornbyi.* It breeds in the mountains of Chile.

One problem facing all birds that spend much of their lives at sea is how to survive storms. The large gliding petrels can undoubtedly run before the wind and are only endangered when blown close to the shore. Storm petrels seem to survive by hugging the surface and keeping to the windward side of the waves. This may seem rather surprising, but on this side of the wave there is an upward air current that assists their flight. The danger period for storm petrels appears to be when the wind changes and runs parallel to the swell. They are then exposed to the full force of the wind without any assistance to keep them airborne except their own strength.

## Two ways of feeding

Like other petrels, storm petrels feed on small fish and crustaceans picked from the surface of the water or caught by diving, and on floating

## EUROPEAN STORM PETREL

CLASS **Aves**

ORDER **Procellariiformes**

FAMILY **Hydrobatidae**

GENUS AND SPECIES *Hydrobates pelagicus*

WEIGHT
**⁸⁄₁₀–1 oz. (23–30 g)**

LENGTH
**Head to tail: 5½–7 in. (14–18 cm)**

DISTINCTIVE FEATURES
**Small size; squared-off tail; pointed wings; black plumage; white rump; white bar on underwings; yellow webs between toes**

DIET
**Mainly surface crustaceans, small fish and cephalopods (squid, cuttlefish and octopuses)**

BREEDING
**Age at first breeding: 4–5 years; breeding season: May–June in south of breeding range, June–August farther north; number of eggs: 1; incubation period: 38–50 days; fledging period: 56–86 days; breeding interval: 1 year**

LIFE SPAN
**Up to 20 years**

HABITAT
**Open seas and oceans; nests on sheltered, undisturbed islands**

DISTRIBUTION
**Breeding: eastern Atlantic and Mediterranean; main colonies in Ireland, Britain and France, with smaller colonies in Iceland, Faeroe Islands, Norway and Mediterranean. Winter: western coasts of Africa, especially off southwestern Africa.**

STATUS
**Common or locally abundant**

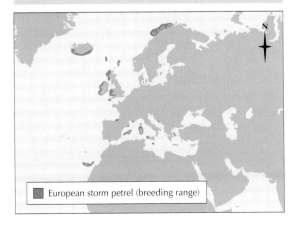

European storm petrel (breeding range)

scraps left by fishers, whalers or predatory animals. The two groups of storm petrels have rather different feeding habits. The long-winged storm petrels skim ternlike over the surface of the water, whereas the round-winged, long-legged storm petrels, such as Wilson's storm petrel, walk on the water with their legs hanging down and wings outstretched while they pick up small animals.

*Some storm petrels, such as Wilson's storm petrel (above), patter the sea's surface with their feet when searching for food.*

### Large egg, slow development

The breeding habits of storm petrels follow the same pattern as those of other petrels. The nest is an unlined burrow in soil or in a crevice between rocks, only rarely being built in the open, against a rock. Storm petrels nest in colonies. About 200,000 pairs of the Galapagos storm petrel nest in one colony in a lava bed on Tower Island.

The female lays a single white egg. It is proportionately huge, being up to a quarter of the adult's weight, and is incubated for 6 weeks, a very long period. The parents share the incubation, taking 2–5 day stints in turn. The chick is brooded for the first 7 days. A large reserve of fat allows the chick to survive periods when there is a shortage of food. At its peak it may weigh more than half again as much as its parents. In the case of Wilson's storm petrel, its fat reserves help it to survive when the burrow is blocked by snow and the parents cannot find it. The chick fledges at 8–12 weeks old. The parents' feeding visits gradually become less frequent and there may be an interval of up to 7 days between the last feed and fledging in the European storm petrel.

# STURGEON

*Using sensitive, whiskerlike barbels under its mouth, a sturgeon feels for prey, chiefly a range of invertebrates and small fish.*

THE STURGEON IS BEST KNOWN as the fish that gives caviar, the luxury food that could soon be a thing of the past. Of greater interest is the fact that the 26 species are relics of primitive fish. They are more or less halfway between the sharks and the bony fish, having a skeleton partly of bone and partly of cartilage. They are sharklike in shape and in the way the hind part of the body turns upward into the upper lobe of the tail fin. The snout is tapered in the young fish, long and broad in adults, and in front of the mouth, on the underside of the head, are four barbels. The body is scaleless except for five rows of large, platelike and sharp-pointed scales extending from behind the gill covers to the tail fin.

## Freshwater giants

The largest species is the Russian sturgeon or beluga, *Huso huso*, of the Caspian and Black Seas and the Volga, Don, Dnieper and other rivers of that region. It reaches 26 feet (8 m) in length and 3,210 pounds (1.5 tonnes) in weight. One specimen, measuring 13 feet (4 m) and weighing 2,200 pounds (1 tonne), was known to be 75 years old. It yielded 400 pounds (180 kg) of caviar. The Atlantic sturgeon, *Acipenser sturio*, found on both sides of the North Atlantic, reaches 11 feet (3.4 m) and 600 pounds (273 kg). The white sturgeon, *A. transmontanus*, of the Pacific coast of North America usually weighs less than 300 pounds (136 kg), but there are records of up to 1,900 pounds (865 kg). The

sterlet, *A. ruthenus*, of the rivers of the former Soviet Union is up to 3 feet (90 cm) long. All other sturgeon species are found in temperate waters throughout the Northern Hemisphere.

## Numbers down everywhere

Sturgeon are slow-moving fish, spending their time grubbing on the bottom for food. Some, however, make long migrations. Individuals tagged in North American waters have been found to travel 900 miles (1440 km). Most species live in the sea and go back up the rivers to spawn. The beluga, from which half the world's supply of caviar comes, is entirely freshwater.

Today all sturgeon are fewer in number than they were a century or two ago: partly from overfishing, partly from the pollution of rivers and to some extent because hydroelectric schemes have spoiled their spawning runs. In the 17th century a prosperous sturgeon fishery flourished in the New England states of North America. In the mid-19th century they were still being caught for their caviar and for a high-quality lamp oil their flesh yielded. A century later the annual catch had fallen by 90 percent.

Sturgeon were at one time abundant off the Atlantic coast of Europe. Now they are found mainly around the mouth of the Gironde River in western France, the Guadalquivir in Spain and Lake Ladoga in the former Soviet Union. Only a few are caught each year around the British Isles and adjacent seas. Around the Black Sea–Caspian Sea area, overfishing has brought the sturgeon yield to a low ebb, and efforts have been made to establish hatcheries to rear young sturgeon and so replenish the stock. It has been estimated that as many as 15,000 sturgeon have been caught in these seas and adjoining rivers in a day. No sturgeon now reach the maximum sizes recorded for their species, and the monsters of yesterday have passed into legend.

## The mud grubber

The name of this fish in several European languages means the stirrer, from the way the sturgeon rummages in the mud for food. This it finds largely by touch, using its sensitive barbels. Sturgeon also have taste buds. In other fish these are normally are on the tongue or inside the mouth, but in the sturgeon they are on the outside of the mouth. The taste buds help in the selection of food. They protrude from the toothless mouth to suck in morsels. Sturgeon are slow feeders and can survive several weeks without eating. In fresh water they eat insect larvae, worms, crayfish, snails and other small fish. In

## WHITE STURGEON

| | |
|---|---|
| CLASS | **Osteichthyes** |
| ORDER | **Acipenseriformes** |
| FAMILY | **Acipenseridae** |

GENUS AND SPECIES **White sturgeon,**
***Acipenser transmontanus***

WEIGHT
**Up to 1,800 lb. (816 kg)**

LENGTH
**Up to 19½ ft. (6 m)**

DISTINCTIVE FEATURES
**Long, gray body; trowel-like snout; 5 rows of bony plates along flanks; curved spines on dorsal ridge; long, forked tail**

DIET
**Bottom-feeder taking shrimps, mollusks, and other crustaceans in sea, and insect larvae, snails, worms and small fish in fresh water**

BREEDING
**Age at first breeding: 10 years; breeding season: spring; number of eggs: 2 million to 3 million; breeding interval: 4–11 years**

LIFE SPAN
**Up to 100 years**

HABITAT
**Coastal marine waters; spawns in rivers**

DISTRIBUTION
**Pacific Coast of North America; landlocked in Columbia River drainage**

STATUS
**Vulnerable: Canada; endangered: U.S.**

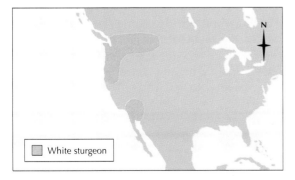

White sturgeon

the sea they take bivalve mollusks, shrimps and other small crustaceans, worms and more small fish than are eaten in fresh water. The beluga feeds in winter mainly on flounder, mullet and gobies in the Black Sea, and on roach, herring and gobies in the Caspian.

Spawning takes place in depths of 18–20 feet (5.5–6 m). The eggs are blackish, 2.5 millimeters in diameter and sticky, so they adhere to water plants and stones or clump together in masses. A single female may lay 2–3 million eggs in a single season. They hatch in 3–7 days, the larvae being ½ inch (12 mm) long at first; during their first summer they may grow to 8 inches (20 cm).

## Use for the swim bladder

In addition to their flesh, oil and caviar, sturgeons have also been fished for isinglass. This substance from the swim bladder was named *huisenblas* by the Dutch in 1525, and this became anglicized to isinglass. When prepared for use, it resembles a translucent sheet of plastic and is almost pure gelatin. It has been used in clearing white wines, an ounce (28.4 g) being enough to clarify up to 300 gallons (4.6 l).

## Royal fish

In the days of Ancient Rome the sturgeon, garlanded with flowers, was piped into the banquet carried by slaves. Edward II of England made it a royal fish. "The King," he decreed, "shall have the wreck of the sea throughout the realm, whales and great sturgeon, except in certain places privileged by the King." At one time any sturgeon caught in the River Thames above London Bridge belonged to the Lord Mayor of London. Henry I is said to have banned even that. Indeed, he forbade the eating of sturgeon at any table than his own.

*An Atlantic sturgeon displays the rows of scutes (bony plates) in its flanks. The adult spends most of its life at sea, returning to rivers to spawn.*

# SUGARBIRD

**With its tubular tongue and long, decurved bill, the sugarbird is adapted to feeding on flower nectar. It probably also helps pollinate Protea species by taking pollen from one plant to another as it feeds.**

THE SUGARBIRDS OF AFRICA were once thought to be related to the honeyeaters of Australasia (discussed elsewhere), although today the two groups are classified in separate families. It is likely that sugarbirds have come to resemble honeyeaters through convergent evolution, both groups having similar feeding habits.

The two species of sugarbirds are dull brown, apart from a yellow patch under the tail. The male grows to about 17⅓ inches (44 cm), of which up to 12 inches (30 cm) is long, flowing tail. The female has a rather smaller tail and measures about 11 inches (27.5 cm) overall. The bill is long and slightly down-curved. The Cape sugarbird, *Promerops cafer*, is found in the Cape Province of South Africa, and Gurney's sugarbird, *P. gurneyi*, is found in the Eastern Cape, Kwazulu-Natal, the Transvaal and Zimbabwe.

Other birds are also known as sugarbirds to aviculturalists (those who undertake the raising and care of birds). These include the so-called blue sugarbird, *Dacnis cayana*, a honeycreeper.

## Conspicuous defense

Sugarbirds are restricted to mountain slopes where the well-known African shrubs of the genus *Protea* are the dominant feature of the vegetation, although they also visit commercial nurseries that grow the plants. The only time that sugarbirds leave *Protea* country is when a shortage forces them to search for other sources.

Outside the breeding season, sugarbirds are inconspicuous among the dense undergrowth despite their long tails, but throughout the breeding season the male sugarbird makes himself conspicuous as he defends his territory. He spends much of his time perched at the top of *Protea* bushes singing a song comprised of creaks and clangs. At intervals he performs a special display flight, going from one bush to another with a jerky action that makes the long tail stream and flutter. This advertisement seems to be sufficient to deter other males, because fighting is rare.

## Tubular tongues

Sugarbirds have the long tongues typical of nectar-feeding birds, which can be rolled to form tubes through which nectar from *Protea* flowers is sucked. As with many of the true honeyeaters, nectar makes up only part of the diet. Insects and spiders are also eaten. Some, such as ant lions and beetles, are taken from the ground and others are caught in the air, in the manner of flycatchers. Small invertebrates are also collected from among the flowers and leaves of *Protea* bushes. The insects are carried in the tip of the bill and beaten against a branch before being swallowed. Females are particularly likely to feed on insects and spiders during the breeding season, when they need sufficient supplies of protein to form eggs and for nourishing the young.

## Two broods

The sugarbird breeds during the South African winter, the same time as the flowering season of the *Protea*, when food is most abundant. The female builds the nest, pressing and molding it into shape with her body. The nest is usually located in a *Protea* bush. It is made of fine twigs, usually of heather, and dry grass molded into a ragged cup 6–7 inches (15–17.5 cm) across. It is lined with the brown fluff from *Protea* flowers and takes about 1 week to build. The female

# CAPE SUGARBIRD

CLASS **Aves**

ORDER **Passeriformes**

FAMILY **Meliphagidae**

GENUS AND SPECIES **_Promerops cafer_**

WEIGHT
**About 1¾ oz. (50 g)**

LENGTH
**Male: 13⅔–17⅓ in. (34–44 cm);
female: 10–11⅔ in. (25–29 cm)**

DISTINCTIVE FEATURES
**Russet upperparts and tail; white throat;
rufous-brown upper breast; streaked lower
breast and belly; yellow vent area; white
area below eyes; brown malar (cheek)
stripes. Male: very long, wispy tail, perhaps
more than 65 percent of total length; long,
decurved bill. Female: long tail, about one-
third length of male's; bill not so long or
decurved as male's.**

DIET
**Mainly nectar; also insects, spiders and
other invertebrates**

BREEDING
**Age at first breeding: 1 year; breeding
season: eggs laid October–January; number
of eggs: 2; incubation period: 17 days;
fledging period: 18–20 days; breeding
interval: 2 broods per year**

LIFE SPAN
**Not known**

HABITAT
**Stands of flowering _Protea_ shrubs on
mountain slopes; commercial _Protea_ nurseries**

DISTRIBUTION
**Cape Province, South Africa**

STATUS
**Common**

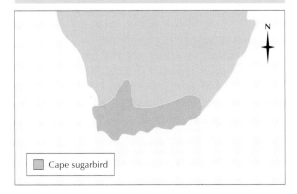

Cape sugarbird

incubates the eggs while the male does no more than defend the territory. The two grayish brown eggs are incubated for 17 days and the chicks are brooded for 6 days all the time and then for a further 2 weeks they are covered only at night.

The male helps feed the chicks but is not as industrious as the female. The chicks are fed on the same food that the adults eat, but they receive more nectar when they are very young. They leave the nest after 18–20 days and are fed for a further 3 weeks, during which time the parents probably have started on the second brood. The male and female remain paired for the whole breeding season.

## Dependence on one plant

There can be few birds that are so dependent on one plant for so many of their needs as the sugarbird. The basis for this dependence lies in the use of _Protea_ shrubs as a source of food. This in itself is unusual, particularly because much of the food consists of insects living on the bushes, which presumably could be collected from other plants. From the dependence on _Protea_ for food it is only a short step to nesting in flowering _Protea_ bushes. While the young are in the nest, the fecal sacs are removed by the parents, as they are in many other birds, but the sugarbird deposits the sacs on particular _Protea_ bushes.

Sugarbirds also use _Protea_ plants as a bath. After rain, they bathe by perching on a stem and flapping their wings against the leaves to send up a shower of water, and also by rubbing their heads against the wet leaves.

_The lifestyle of the
sugarbird is very closely
linked to_ Protea _shrubs.
The breeding season,
during the South
African winter, peaks
when certain_ Protea
_species are flowering in
the lowlands._

# SUMATRAN RABBIT

*The Sumatran rabbit's brown striped coat is unique among rabbits.*

THE SUMATRAN RABBIT, also known as the short-eared rabbit, is found only on the island of Sumatra in Indonesia, and may well be the world's rarest rabbit. Apart from one confirmed sighting in the Guning Leuser National Park in northwest Sumatra, and a sighting in 1997 in Mount Kerinci National Park, central Sumatra, the only record of the species dates back to 1916. No living specimen has been seen or studied by zoologists since the 1930s.

It is rather small, only about 14–16½ inches (36–42 cm) long in head and body, with a short inconspicuous tail and remarkably short ears It has a soft, dense underfur with longer, harsher hairs on the outside. The Sumatran rabbit's coat has a highly distinctive coloring. Its ground color is the usual rabbit or hare color of buff gray, but it is prominently marked with long, irregular bands of dark brown. One of these bands runs down the middle of the back from the snout to the tail, another broad one curves from the shoulder to the rump and from the rump down to the hind leg, while yet another narrow curve runs from the shoulder halfway down the upper foreleg. The rump and tail are bright red and the limbs are grayish brown. The underside of the neck is dark brown, and the remaining underparts are buff white.

Scientists believe that the Sumatran rabbit has never been abundant in any part of its range. Today, it is restricted to the tropical forested areas of the Barisian Mountains in west and southwestern Sumatra.

## Shy and antisocial

The Sumatran rabbit is apparently a nocturnal animal, sheltering by day in burrows on the forest floor. It does not dig these for itself but uses burrows made by other animals.

The Sumatran rabbit differs from other rabbit species in several respects, apart from its unique coloration and patterning. For example, unlike other rabbits it does not form large colonies, preferring to live in pairs. It is a shy animal but, when disturbed, it does not have the speed or quickness of movement of hares. Because of its extreme rarity, the Sumatran rabbit's breeding behavior remains a matter of scientific debate.

# SUMATRAN RABBIT

| | |
|---|---|
| CLASS | **Mammalia** |
| ORDER | **Lagomorpha** |
| FAMILY | **Leporidae** |
| GENUS AND SPECIES | ***Nesolagus netscheri*** |

ALTERNATIVE NAMES
**Sumatran hare; short-eared rabbit;
Sumatran short-eared rabbit; Kerinci
rabbit; Kelinci kerinci**

WEIGHT
**3⅓–6⅔ lb. (1.5–3 kg)**

LENGTH
**Head and body: 14–16½ in. (36–42 cm);
tail: ½–⅓ in. (1.5–1.7 cm)**

DISTINCTIVE FEATURES
**Soft, dense underfur, with coarser guard
hairs; gray-brown upperparts; brown stripes
along back from nose to tail and on face,
legs and body; bright red rump and tail;
dark brown throat with white underside;
very small ears**

DIET
**Stalks and leaves of ground-level
forest vegetation**

BREEDING
**No details known**

LIFE SPAN
**Up to 1 year in captivity**

HABITAT
**Tropical rain forest, in highlands up to
altitude of 7,590 ft. (2,300 m)**

DISTRIBUTION
**Mountain ranges in western and
southwestern Sumatra**

STATUS
**Critically endangered**

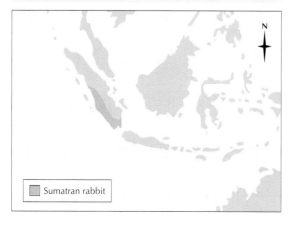

Sumatran rabbit

The Sumatran rabbit feeds on young shoots and the stalks of leaves of forest undergrowth plants and sometimes damages the trees by gnawing away the bark. It has been kept in captivity for periods of up to a year, during which time it consumed a fairly varied diet of cooked rice, young maize and bread. The rabbit took fruit such as ripe bananas and pineapple but, perhaps surprisingly, refused cultivated vegetables, roots and the bark from various trees.

As scientists have had so few opportunities to study the Sumatran rabbit in the wild, information about its predators remains mostly speculative. However, it is likely that the animal is preyed upon by snakes and smaller cat species.

## Ecological enigma

There are two distinctive points of interest concerning the Sumatran rabbit. The first is that it should be so localized and so rare. The second point is that it has such small ears. These two factors may not be wholly unrelated. It is a general scientific rule that there is a shortening of the extremities in animals, especially of the ears, the nearer to the poles an animal's distribution gets. The jack rabbits, genus *Lepus* (discussed elsewhere), which live in southern North America have very long ears, which facilitate heat loss and thus enable the animals to remain cool in high temperatures. The ears become shorter and shorter for successive species the farther north they range, as heat conservation is essential in colder climates and this necessitates a minimum of unnecessary surface area. In the snowshoe rabbit, *L. americanus*, the most northerly rabbit species, the ears are quite small.

However, in apparent contradiction to this tendency, the Sumatran rabbit lives almost on the equator and yet has unusually small ears. This may be because it lives in dense forests in perpetual shade, where the temperature is always cool. This also may explain why the species is so rare: it could not leave this specialized habitat without growing longer ears.

## The world's rarest rabbit?

The Sumatran rabbit's forest environment has come increasingly under pressure as land has been taken over for the development of tea and coffee plantations. Consequently, the population of this already extremely rare rabbit has suffered, and until recently scientists feared that it had become extinct. However, in November 1997 a team from the organization Flora and Fauna International found a specimen in the Mount Kerinci National Park. Using an automatic camera trap, the team managed to take the first-ever photograph of a Sumatran rabbit, on a mountain range at an altitude of 7,590 feet (2,300 m).

# SUN BEAR

A full-grown male sun bear may weigh as much as 150–200 pounds (68–91 kg); however, the more usual weight range is 59–143 pounds (27–65 kg). The sun bear is a stocky animal with short, bandy legs. It also has large paws with long, strong claws. Its feet have naked soles, unlike bears that live most of the time on the ground and have hairy soles. The muzzle is shortened and the ears are small and rounded.

The smooth, sleek coat of coarse, short hair is uniformly black except for the grayish yellow muzzle and the distinctive crescent or U-shaped mark on the chest, which may be creamy or yellowish or any shade in between. This mark varies in form and may sometimes be lacking. Occasionally, the feet are light colored.

The sun bear lives in the tropical and subtropical forests of southern Asia, in Myanmar (Burma), south through Southeast Asia and Peninsular Malaysia, to the islands of Sumatra and Borneo. Its range reaches southern China.

## Tree climber

The sun bear is a skilled climber, spending most of its time in the tops of tall trees. It is active mainly at night. By day it sleeps or sunbathes in a nest formed of branches and twigs. The nest is situated at the top of a tree, sometimes as high as 23 feet (7 m) above the ground. Unlike some of the bears in temperate climates, the sun bear never hibernates.

Because of the inaccessibility of its habitat, the sun bear is difficult to observe in the wild, especially as it is very cautious and wary. In captivity it has made an intelligent and lively pet

*The sun bear's front feet have five long claws for tearing open trees and other insect sites. The bear also has a long tongue that helps it reach insects and honey.*

THE SUN BEAR IS traditionally thought to have derived its name from the crescent-shaped mark on its chest, due to the fact that the yellow crescent is held to represent the rising sun in Eastern folklore. It is also called the honey bear, the Malaysian bear and the bruang.

The sun bear is one of the smallest of all bear species, along with the Asiatic black bear, *Ursus thibetanus* (discussed elsewhere), and spends most of its time in trees. It has a head and body length of about 39–55 inches (100–140 cm) with a tail 1⅛–2¾ inches (3–7 cm) long, and stands only about 27½ inches (70 cm) at the shoulder.

when young, but after a few years it frequently becomes dangerous. The sun bear is regarded as a pest in fruit plantations.

## Honey eater

The sun bear has a very varied diet, feeding on any small rodents, lizards, small birds or insects that it can find. It also eats fruit and the soft growing part of the coconut palm, known as palmite. The sun bear digs out termite nests or ant nests, inserting its forepaws one at a time and then licking the ants or termites off them. It is very fond of honey and tears open trees to find

# SUN BEAR

| | |
|---|---|
| CLASS | **Mammalia** |
| ORDER | **Carnivora** |
| FAMILY | **Ursidae** |
| GENUS AND SPECIES | **Helarctos malayanus** |

**ALTERNATIVE NAMES**
Honey bear; Malaysian bear; bruang

**WEIGHT**
Usually 59–143 lb. (27–65 kg); full-grown male: 150–200 lb. (68–91 kg)

**LENGTH**
Head and body: 39–55 in. (100–140 cm); shoulder height: about 27½ in. (70 cm); tail: 1⅕–2¾ in. (3–7 cm)

**DISTINCTIVE FEATURES**
Stocky build; short legs; large paws with naked soles; blunt, gray or yellow muzzle; short, glossy black coat with variable creamy yellow breast patch; feet may be pale color

**DIET**
Honey, bee larvae and other insects, fruits, tree sap, vegetation, small rodents and birds

**BREEDING**
Age at first breeding: 2–3 years; breeding season: year-round; gestation period: 95 days, capable of delayed implantation; number of young: 1 or 2; breeding interval: up to 2 litters per year

**LIFE SPAN**
Up to 30 years in captivity

**HABITAT**
Dense forest up to high altitudes

**DISTRIBUTION**
Easternmost India east to south-central China and south to Sumatra and Borneo

**STATUS**
Vulnerable; locally common in some areas, rare in others

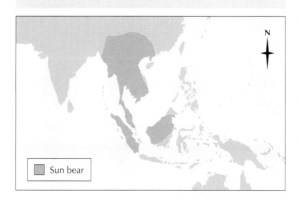

Sun bear

the nests of wild bees, from which it extracts the honey and grubs with its long, extensible tongue. Most bears, including the brown bear, *U. arctos* (discussed elsewhere), are able to push the tongue out to a surprising length even when they are merely yawning.

## Two cubs in a litter

There is still some scientific uncertainty about the breeding habits of the sun bear. However, it is known to have a variable gestation period and to be capable of delayed implantation. In view of the latitudes in which it lives, the sun bear probably has no regular breeding season, although births of sun bear cubs have been recorded on several occasions during August. There are usually two cubs in a litter. They are born on the ground and well hidden in the vegetation on the forest floor. They remain with their parents for some time after their birth.

*Excellent climbers, sun bears spend considerable time in trees. Without an adequate range they are often forced to raid agricultural crops and gardens for food.*

## From cub to bear

One of the few detailed accounts of the growth of a sun bear was given by James Alexander Hislop in his book *Natural History*, published in 1955. A tiny sun bear cub was adopted by Hislop and his family after it apparently was abandoned by its mother in Peninsular Malaysia at the age of two weeks old. The family named it Bertie.

For the first few weeks, the cub was effectively blind and would scramble about on the floor, bumping into anything in its way. It took easily to bottle-feeding and while sucking its bottle made noises rather like a duck grubbing in the mud. It was extremely fond of bread spread with honey, syrup or jam and would come and beg for it in the evening. When contented, the cub would sit for hours, sucking a hind paw while making a humming sound, in the manner of a human baby sucking its thumb before going to sleep. After sucking its paw, the cub would fall asleep, always on its belly, head tucked underneath its forepaws.

There are several accounts of hand-reared bear cubs of other species, and all indicate that in many ways bear cubs are very like human babies. One outstanding similarity is that when feeding or resting cubs are apt to develop mannerisms or types of behavior that they persistently carry out. Then, suddenly, they abandon them. One bear cub, for example, insisted for a short period on rhythmically swinging one leg while taking its bottle, refusing to feed if unable to do this.

At six months old Bertie was taken to a national park so that he might, in time, return to the jungle nearby, but he would never wander for more than two days and always came back to his enclosure.

In the park he soon became an adept tree-climber and made platforms of twigs and branches in the same way that the sun bears in the wild make their sleeping nests. He discovered ants and other insects in decayed tree stumps and logs, and he used his powerful claws to demolish the rotten wood, while his long tongue lapped up the ants.

At the age of nearly two years, Bertie weighed more than 120 pounds (54.5 kg) and measured 46 inches (115 cm) from head to tail. From then on his size and weight became too great for him to be kept as a pet, and his owners were subsequently compelled to send him to a zoo.

*The sun bear's habitat is declining because of human activities, and the animal could become extinct in the wild within the next few decades. This bear is pictured in San Diego Zoo, California.*

# SUNBIRD

SUNBIRDS ARE THE Old World equivalent of the American hummingbirds. They are not related, however, and sunbirds' wingbeats are not as rapid as those of hummingbirds, but they do both feed on nectar and small insects. The main similarity, apart from diet, is in the plumage. That of the males generally has a metallic luster or velvety sheen. Many males and some females have epaulettes of bright feathers on the shoulders that are raised during displays.

The 130 sunbird species are all small, the largest measuring only 8 inches (20 cm) from bill to tail. Most are less than 6 inches (15 cm) long. The sunbird family ranges across most of the warmer forested areas of the Old World. About half the species are found in tropical Africa. The rest live in southern Asia, from Israel through India and Indonesia to northern Australia and the Philippines. Sunbirds are also found on the Seychelles and on the islands of Aldabra and Assumption in the Indian Ocean. Some sunbirds in Southeast Asia are called spiderhunters, but this name does not seem to be particularly related to their feeding habits.

## Following their food
Sunbirds live in forests, bushy cultivated land and savanna, although they have also been found well up mountains near the timberline. They also frequent towns and gardens. Here they have been found roosting, sometimes communally, in holes in trees or abandoned nests, apparently in order to keep warm during the cool nights.

Sunbirds feed on insects and nectar. They fly about in parties, but are not truly gregarious; the party does not stay together as a social unit. Instead, the sunbirds come together at plentiful sources of food and move around the country as different flowers come into season. However, some insect-eating sunbirds have been seen banding into parties and working through the vegetation together, driving insects before them. Some sunbirds follow a regular track in search of food supplies. One example is the pygmy sunbird, *Anthreptes platurus*, which moves south from the Sudan to the Democratic Republic of Congo (Zaire), staying there for a few months. While it is there, the plants of the savanna flower and the sunbird raises a brood, although it depends less on flowers than its longer-billed relatives do.

## Diet of nectar and insects
Sunbirds resemble flowerpeckers and honeyeaters (both discussed elsewhere) in their feeding habits and the specialized structures adapted for them. For example, all three groups feed on nectar, insects or both. They have special tongues to get nectar from the bases of long, thin flowers and a digestive system that enables large quantities of liquid to be passed through rapidly.

The tongue is divided into three or four flaps at the tip, which help spoon up the nectar. For most of its length the tongue is rolled over on each side to make two tubes through which the nectar is sucked. The muscular gizzard, which is used to crush hard-bodied insects, has an exit and entrance near each other so it can be shut off from the rest of the digestive system, allowing nectar to run straight into the intestine from the gullet.

In order to obtain nectar a sunbird usually perches on a twig next to a flower and probes inside it with its long, curved bill. If the flower is too large for the sunbird to reach the nectar, it tears a hole in the side of the flower near the base and pokes its bill in that way. Some flowers have springlike filaments that release a cloud of pollen onto the sunbird's forehead when triggered by the action of the bird's bill. Later, this pollen is transferred to the stigma of another plant, thereby fertilizing it. Among the plants pollinated this way are the tropical mistletoes. These plants rely on sunbirds, flowerpeckers and honeyeaters for pollination and would die out without the birds' unknowing assistance.

*It is likely that sunbirds originally fed on insects living in flowers. The flowers then evolved nectar supplies to attract the birds. This is a female black sunbird,* Nectarinia amesthystina.

*Unlike hummingbirds, sunbirds prefer to perch when feeding, rather than hover. The photograph shows a male collared sunbird,* Anthreptes collaris, *on an aloe flower.*

## Aggressive territory battles

Male sunbirds are aggressive in the breeding season, chasing intruding males well away from their territories. Some males have a strong song; others make only a twitter. They usually do not help the females build the nests or feed the young, yet the pair often stays together year-round.

The nest is a baglike structure of roots and grasses matted and woven together, suspended from a twig. There is an entrance on one side, sometimes covered with a porch. Each species has its own variations in design and building materials. The purple sunbird, *Nectarinia asiatica,* for example, which ranges across India, covers its nest with small pieces of bark, caterpillar droppings and bits of paper, string and other trash. The nest is often suspended from a bare branch with no attempt at concealment. Some species nest near the nests of wasps, apparently gaining protection from their presence. Most spiderhunters make a cup-shaped nest that is sewn to the underside of a leaf by fibers or cobwebs and knotted on the upperside. The entrance to this kind of nest is in the side of the cup.

The female broods the one or two eggs for two weeks and the chicks stay in the nest for 2–3 weeks before traveling about in family parties.

## Sunbirds unlock flowers

The flower of some African mistletoes will not open without the aid of a sunbird. In one species of mistletoe, the flowers are opened by the sunbird inserting its bill into a slit in the side of the corolla. The tube then bursts open and the anthers spring out, scattering pollen onto the bird's feathers. In another species, the anthers snap off and fly out. The flower of one Indian species has to be squeezed gently at the tip by a sunbird before it will open.

## PYGMY SUNBIRD

| | |
|---|---|
| CLASS | **Aves** |
| ORDER | **Passeriformes** |
| FAMILY | **Nectariniidae** |
| GENUS AND SPECIES | ***Anthreptes platurus*** |

**WEIGHT**
⅕–¼ **oz. (5.2–7.3 g)**

**LENGTH**
**Head to tail: male, 6½ in. (16.5 cm); female, 2½ in. (10 cm)**

**DISTINCTIVE FEATURES**
**Slender body; long, slightly decurved bill. Male: olive-green head, breast and upperparts; bright yellow belly and undertail; very long central tail streamers. Female: duller greenish gray upperparts; yellow underparts. Both sexes: long, slightly decurved bill.**

**DIET**
**Invertebrates and flower nectar**

**BREEDING**
**Age at first breeding: 1 year; breeding season: December–February (Nigeria); number of eggs: 1 or 2; incubation period: about 14 days; fledging period: 12–15 days; breeding interval: 2 broods per year**

**LIFE SPAN**
**Not known**

**HABITAT**
**Savanna, particularly acacia scrub; also open woodland and gardens**

**DISTRIBUTION**
**Sub-Saharan Africa in savanna zone, from Senegal east to Uganda and north to northern Chad**

**STATUS**
**Common**

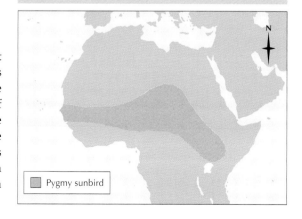

Pygmy sunbird

# SUN BITTERN

THE SUN BITTERN IS A large and little-known inhabitant of tropical American forests. Recent research, based on analysis of egg-white proteins, suggests the kagu, *Rhynochetus jubatus*, of Australia is its closest relative, not the true bitterns of the heron family Ardeidae. Both birds may have a common ancestor that originated in the Mesozoic Gondwanaland (the original landmass from which Earth's continents originated), with the subsequent drifting apart of Australia and South America allowing for evolutionary divergence.

The sun bittern has a heronlike appearance, and is 17–19 inches (43–48 cm) in length, with a long, slender neck, small head and long bill. The bright orange legs are also long and slender and the toes are unwebbed. The wings and tail are broad. The feathers are soft and are mainly brown and gray with black bars and spots. The crown is black and two white streaks run across the face. There are two broad black bands across the tail. The bill is black on the upper mandible and yellow on the lower mandible. When a sun bittern opens its wings, a striking pattern of chestnut and orange becomes visible on the back, with white and black patches on the wings.

## Sunset display

Sun bitterns range from Guatemala south to Bolivia and central Brazil. Like herons, they live singly or in pairs along the banks of rivers or in swampy woodland and wade slowly through the shallows in search of food. Captive sun bitterns have been described as standing with their bodies swaying from side to side in the same manner as bitterns, reputedly to make themselves less conspicuous among the waving reeds. They also spend a considerable time motionless, the neck withdrawn in the manner of herons. Sun bitterns are reluctant to fly, preferring to walk and to swim across streams. When disturbed, however, they fly into trees. Their flight is very quiet because of their soft plumage. Sun bitterns usually are silent but may utter quiet whistles or rattles or ascending or descending trills. Even 6-day-old chicks exchange trills with attending adults.

The defensive display of the sun bittern is most spectacular. The bird lowers the forepart of its body, lifts its head and spreads its wings with the rear edges raised. It lifts the tail and brings it up so that the strikingly patterned plumage is displayed in a semicircle. During the display the bird makes a harsh rattling sound.

At one time ornithologists believed this display was related to courtship, but now it is thought to be defensive. Courtship display may involve the bird making flight displays 33–49½ feet (10–15 m) high, just over the tops of the trees. This is followed by a sharp *kak-kak-kak* call, ending in a trill while the sun bittern glides down with wings outstretched and hanging from the body, to display its colorful pattern.

## Nests rarely seen

The nest is constructed by both sexes on a fairly horizontal tree branch 3⅓–23 feet (1–7 m) above the ground. It consists of a slightly oval cup about 8 inches (20 cm) by 6½ inches (16 cm). The base is made of long grass fibers, wrapped into an oval shape and attached to the branch with mud. On this base the birds build up the sides with leaves, rootlets, moss and mud. Seeds may sprout in the mud, helping to camouflage the nest.

The first record of the nesting behavior of the sun bittern was a description of a pair that nested in London Zoo in 1865. It is still the most detailed account, although sun bitterns have since nested in other zoos. The pair built their nest of straw, grass, mud and clay on a specially provided platform 10 feet (3 m) up. The first egg was found broken under the nest; a second was

*The sun bittern feeds on insects, crustaceans, small fish and other small animals found in shallow water along the banks. It catches its prey with heronlike jabs of its bill.*

*This adult sun bittern is performing a defensive display in order to defend its chicks.*

## SUN BITTERN

| CLASS | Aves |
| --- | --- |
| ORDER | Gruiformes |
| FAMILY | Eurypygidae |
| GENUS AND SPECIES | *Eurypyga helias* |

WEIGHT
**About 6¾ oz. (190 g)**

LENGTH
**Head to tail: 17–19 in. (43–48 cm)**

DISTINCTIVE FEATURES
**Slim, elongated body; long bill and neck; fairly long tail and legs; cryptically colored (coloration provides camouflage); 2 white lines across blackish head; brownish upperparts with fine black and gray markings; white throat and vent; orange-yellow legs; slight differences between 3 subspecies**

DIET
**Invertebrates, including crustaceans and insects; vertebrates, including fish, frogs and lizards**

BREEDING
**Age at first breeding: 2 years (in captivity); breeding season: eggs laid March–May, at start of wet season; number of eggs: 2; incubation period: about 30 days; fledging period: about 22–30 days; breeding interval: about 1 year**

LIFE SPAN
**More than 30 years in captivity**

HABITAT
**Humid forest with open understory, near water such as rivers or swampy areas**

DISTRIBUTION
**Guatemala south through Central America and south along west of South America to Ecuador; also, Amazonian South America to altitude of 6,100 ft. (1,850 m)**

STATUS
**Fairly common in some areas, rare in others**

laid shortly afterward and was incubated by both parents for 27 days. The chick was like that of a snipe and was fed by both parents on food carried in their bills until its wing feathers had grown enough for it to fly to the ground at the age of 21 days. The parents continued to feed it, and 2 months after it had hatched another egg was laid and incubated mainly by the male while the female continued to feed the first chick. In the wild, the usual clutch seems to be two eggs.

### Diverse order

The order Gruiformes, to which the sun bittern belongs, contains some unusual birds. These include the large family of rails, some of which are flightless; the button quails, in which the female plays the leading role in courtship; the mesites of Madagascar, which probably are flightless; and the cranes, finfoots and bustards (all discussed elsewhere in this encyclopedia). Some resemble birds outside the order, such as the storklike kagu, the ibislike limpkin, *Aramus guarauna*, and the heronlike sun bittern.

Despite a variety of external form and habits, the gruiform birds have many similarities in the form of their skeletons and muscles. One habit that is very common in the group, however, is that of nesting on the ground and producing chicks that can walk soon after hatching. The sun bittern is an exception because it nests in trees and, although its chicks are hatched with a coat of down and appear well developed, they are fed in the nest for some time.

Sun bittern

# SUNFISH

THERE ARE TWO KINDS of sunfish. One, the species *Mola mola*, is a marine giant that gets its name from the mistaken idea that it rises to the surface to bask in the sun; it is covered under a separate entry. The other, covered here, is a group of freshwater fish the behavior of which is influenced by the sun.

Freshwater sunfish are North American. The 27 perchlike species in the family Centrarchidae are variously called bass, crappies and blue gills, as well as sunfish. They usually have a long, continuous dorsal fin, the fore part of which is spiny. In a few species there is a slight notch where the fore and hind parts meet, the latter part being higher.

Several species have a so-called ear flap on each side of the head, where the operculum (gill cover) extends backward. This is often made more obvious by its coloring: white-edged in the long-eared sunfish, *Lepomis megalotis*, sometimes with dots of red on the flap itself, or with a blood-red blotch as in the pumpkinseed or common sunfish, *L. gibbosus*. The blue spotted sunfish, *Enneacanthus gloriosus*, is 3½ inches (9 cm) long. The largemouth black bass, *Micropterus salmoides*, may be 30 inches (75 cm) long and weighs up to 25 pounds (11.4 kg).

Sunfish are most common in central and eastern regions of the United States. The only species native to the west of the Rockies is the Sacramento perch, *Archoplites interruptus*, although others have now been introduced into the rivers of California.

## A fish-eat-fish world

Most sunfish live in clear lowland rivers and lakes, especially where there is a sandy bottom, and particularly in flowing waters with quiet, weedy shallows. They feed on swimming animals, rarely touching prey on the bed. The smaller species roam in small shoals, eating insect larvae, especially midge larvae, and small crustaceans. The larger sunfish species are solitary predators, eating other fish.

## Sensitive kind

All sunfish are sensitive to changes in their environment, notably to sudden changes of temperature. For example, the red-breast sunfish, *Lepomis auritus*, winters in the deeper water of lakes and migrates into shallows to spawn

when the water temperature is 50° F (10° C). It is, however, strongly influenced by sunlight, possibly largely through its effect on temperature. On a dull day the fish is fairly inactive. Usually, a male sunfish's first act on reaching a spawning ground is to dig a redd, a shallow depression in the sand, using his tail as a fan. On dull days, however, he does no more than station himself over the redd, fanning it with his fins to prevent it from silting up. On a sunny day the shadow from a passing cloud is enough to make him cease all activities.

## Male guards nest

The redd is usually in a spot sheltered by water plants or large stones, and the male may use plant pieces to reinforce the nest; the smaller the fish, the shallower the water covering the nest site. After preparing the site, the male entices a female in to lay her eggs. In some species there is no great difference between male and female, while in others the male is more brightly colored. The female lays about 1,000 eggs, which tend to clump together. They also stick to sand grains, which serve as camouflage. Once the female has

*A mature pumpkinseed among waterweeds. As they mature, sunfish graduate from shoaling and feeding on insect larvae to a more solitary existence and substantial diet.*

*The banded sunfish, Enneacanthus obesus, is common in creeks and quiet backwaters of the eastern United States, from New Hampshire south to Florida.*

laid her eggs, she is driven away by the male, who guards the nest, fanning it with his fins to aerate the eggs and chasing away any intruder.

The eggs hatch in 3–6 days. At first the fry lie in the redd, later clinging to the water plants, but the male remains in attendance for 2–3 weeks, shepherding his tiny charges into the redd each night. The male has good cause to guard his own eggs and young, for other young sunfish will eat them if given the chance.

## Colors and movement

One of the most colorful species is the pumpkin-seed sunfish. The male is grayish green with between five and eight pearly bars on the flanks. As well as the blood-red blotch on the black ear flap, there are reddish or orange spots on the head and orangish red markings on the throat and belly. The operculum itself is green with red lines and spots, and the fins are green to yellow. Tests in an aquarium revealed that males reacted to both color and movement, either courting or showing aggression to another fish depending on its coloration and its movements.

## Cuckoo shiner

A form of cuckoolike behavior is exhibited by the redfin shiner, *Lithrurus umbratilis*, another native of North America. While a male green sunfish establishes his nest, the shiners assemble above him. When a female sunfish lays her eggs, the shiners lay their own in the same nest. The shiners are first attracted by the movements of the male sunfish, but the stimulus that makes the shiners spawn is a chemical in the milt (sperm-containing fluid) or eggs of the sunfish.

# SUNFISH

| CLASS | Osteichthyes |
| --- | --- |
| ORDER | Perciformes |
| FAMILY | Centrarchidae |
| GENUS | 8 genera |
| SPECIES | 27 species, including pumpkinseed, *Lepomis gibbosus* (detailed below) |

WEIGHT
**Up to 22 oz. (630 g)**

LENGTH
**Up to about 16 in. (40 cm)**

DISTINCTIVE FEATURES
**Deep, laterally flattened body; moderately sized head; olive to golden-brown back and flanks; orange belly; dusky bars on flanks; black and red markings on gill covers**

DIET
**Worms, crustaceans and insects; also small fish and other vertebrates**

BREEDING
**In May–June male builds redd (spawning nest) in shallows near shore, where female lays eggs; male guards fry for 11 days or so after they hatch; he then prepares redd for another spawning with same or new female**

LIFE SPAN
**Up to about 9 years**

HABITAT
**Quiet, weedy lakes, ponds and pools of creeks and small, slow-flowing rivers**

DISTRIBUTION
**Northeastern North America, from New Brunswick to South Carolina; introduced to northwestern U.S. and central Europe**

STATUS
**Locally common**

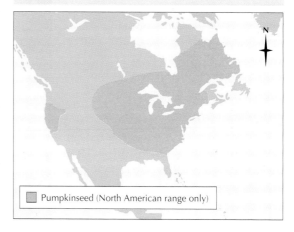

Pumpkinseed (North American range only)

# SUNI

The SUNI, *Neotragus moschatus*, and its relatives the dwarf antelopes are the smallest of all the antelopes and have a slender, dainty shape. They are closely related to the steenbok, *Raphicerus campestris* (discussed elsewhere), and the grysboks, genus *Raphicerus*, but are smaller, with the horns in line with the face instead of upright. In common with the steenbok, dwarf antelopes have a gland in front of the eyes, although it is not in a deep pit. There are also glands between the hooves, but in the steenbok these open into long clefts, while in the dwarf antelopes they open by only small holes. Like their relatives, however, the dwarf antelopes have no knee tufts, and have either small lateral hooves or none.

The suni is 12–16 inches (30–41 cm) tall with strongly ridged horns that are 2½–5 inches (6.3–12.5 cm) long; the females have no horns. There are no lateral hooves. The coat color is reddish to grayish brown and slightly speckled; the underside is white. The suni ranges along the East African coast, from Kwazulu-Natal in the south to the Tana River, Kenya, in the north, and reaching inland to Kariba and Ngorongoro. It also occurs on the islands of Zanzibar and Mafia, but not on nearby Pemba.

Bates' dwarf antelope, *N. batesi*, is smaller than the suni, growing to slightly more than 13 inches (33 cm) high, with conical horns no longer than 2½ inches (6.3 cm). It has small ears and large eyes, extremely slender legs and rudimentary lateral hooves. The coat color is brown, with the underside usually only slightly whiter. Bates' dwarf antelope is found from southeastern Nigeria east to western Uganda.

The smallest of all dwarf antelopes is the royal antelope, *N. pygmaeus*, 10¼ inches (26 cm) high or less, with conical horns only 1½ inches (3.8 cm) long, in the males only. Its hindquarters, as in many small ungulates (hoofed members of the order Artiodactyla), are higher than the forequarters and are extremely muscular, enabling the antelope to perform fast, high jumps. The royal antelope has long, thin legs and a high-stepping gait. The lateral hooves are absent, but there is an area of naked skin where they would be located in a suni. Its coat color is bright reddish fawn, with a white chin, throat, chest and belly. The royal antelope is found along the West African coast from Sierra Leone to Congo.

## Secretive lifestyle

All three species of dwarf antelopes live in dense cover. The royal and Bates' dwarf antelopes live in humid tropical forests, whereas the suni favors drier thicket country interspersed through the eastern savannas. Consequently, the suni's distribution tends to be local and fragmented. There is still some scientific debate as to the suni's lifestyle. Generally it seems to be solitary and is rarely seen in groups of more than two.

These tiny antelopes seem to graze more frequently than they browse, although Bates' dwarf antelope are believed to eat the tops of peanut plants. Suni have been seen to feed on roots and tubers, which they presumably dig up with their sharp hooves. Suni spend the day lying in thick cover, coming out into more open clearings to feed at dawn and dusk. Royal antelope are largely nocturnal animals.

When an intruder approaches a suni, the animal lies hidden, blending well with the background, until the enemy is perhaps only 10 yards (9.1 m) away. It then jumps up and dashes away, twisting and dodging. However, after running for about 300 yards (275 m), the suni stops, stands still and looks back. When a suni hears or smells anything unusual, it gives a short bark and a whistling alarm call.

*In common with other dwarf antelopes, the suni favors thick undergrowth and broken forest. It has favorite sites within this habitat in which it regularly lies up.*

*The suni's naturally pale coloration and the variably tinted patches on its coat provide it with useful camouflage in sun-dappled forest.*

## DWARF ANTELOPES

| | |
|---|---|
| CLASS | **Mammalia** |
| ORDER | **Artiodactyla** |
| FAMILY | **Bovidae** |

GENUS AND SPECIES **Royal antelope, *Neotragus pygmaeus*; Bates' dwarf antelope, *N. batesi*; suni, *N. moschatus***

WEIGHT
**Suni: 8¾–13¼ lb. (4–6 kg)**

LENGTH
**Suni. Head and body: 22½–24½ in. (57–62 cm); shoulder height: 12–16 in. (30–41 cm); tail: 3⅕–5⅕ in. (8–13 cm)**

DISTINCTIVE FEATURES
**Tiny antelope; russet coat; lighter underparts; short, slender horns (male only)**

DIET
**Grasses, leaves, fruits and flowers**

BREEDING
**Age at first breeding: 6–18 months; breeding season: often year-round, but seasonal peaks occur; number of young: 1; gestation period: about 180 days; breeding interval: 1 year**

LIFE SPAN
**Up to 14 years in captivity**

HABITAT
**Brushland, forest clearings and forest**

DISTRIBUTION
**Suni: East African coast, from Kenya south to Kwazulu-Natal; Zanzibar and Mafia Islands. Bates' dwarf antelope: Central African Republic, Democratic Republic of Congo (Zaire) and Uganda. Royal antelope: West African coast, from Sierra Leone east and south to Congo.**

STATUS
**Suni: conservation-dependent; Bates' dwarf antelope and royal antelope: near-threatened**

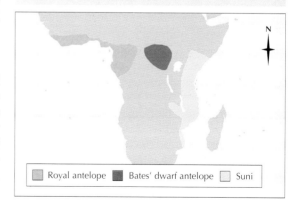

Royal antelope Bates' dwarf antelope Suni

Royal antelope are even more shy than suni and are equally secretive. With their long hind legs, they are powerful jumpers, and can cover a distance of 9½ feet (2.9 m) in a single jump.

Suni live in a seasonal environment and breed at set times of the year. In South Africa the young are born during the month from mid-November to mid-December. The male herds the female, chasing her around, making a bubbling noise and scent-marking their territory. Both suni and Bates' dwarf antelope have scent glands that they use for this purpose. Their smell also denotes the sex and status of the individual and may repel insects.

## Unfamiliar antelope

Dwarf antelopes are among the least known of all ungulates; even specimens in museum collections are rare. In the past, royal antelopes have often been confused with chevrotains, family Tragulidae, which vie with them for the title of smallest ungulate, as well as with the far larger blue duiker, *Cephalophus monticola* (discussed elsewhere in this encyclopedia).

# SURFPERCH

A HUNDRED YEARS AGO the surfperch fish were described as extraordinary not only because they bear fully hatched young but also because the young are sexually mature almost as soon as they emerge from the female.

There are 24 species of surfperch, also called seaperch or surffish. They are very similar in form to the freshwater perch except that the dorsal fin, spiny in front and soft-rayed in the hind part, is continuous rather than in two distinct parts. There is also a groove in the back on each side of the dorsal fin. Their bodies are laterally compressed and their lips are fleshy. They are mainly silvery, darker on the back than on the belly. At the leading edge of the male's anal fin is an intromittent organ for transferring the milt (sperm-containing fluid) to the female. Surfperch are 5–18 inches (12–45 cm) long.

Most surfperch species swim up into the surf to drop their young; the rest of the time they live in shallow seas. The pink seaperch, *Zalembius rosaceus*, is unusual in living at depths of about 600 feet (180 m); others may be found at times in tide pools, and the freshwater tule perch, *Hysterocarpus traski*, lives in the Sacramento River in California. Except for three species living off Japan and Korea, all seawater surfperch live off the Pacific coast of North America, from Alaska to Baja California. Some of the larger species are fished commercially or for sport, both in the United States and farther afield.

## Infantile mating

Surfperch are in every way ordinary, perchlike fish, feeding on small crustaceans and other small invertebrates, except in their breeding, which goes on more or less all summer long. In some species at least, the male is slightly smaller than the female and somewhat darker in color during the winter. The male is sexually mature at birth, the female maturing soon after. Pairing then takes place, usually in shallow water within two days of birth. The young are born during the following summer at the water's surface. This early sexual maturity and breeding behavior is probably unique among vertebrates.

*The midnight surfperch Macolor macularis, also known as a seaperch or snapper, over a coral reef in the Philippines. This is one of just three species found outside the United States.*

## Living in the ovary

Another unusual feature is that the eggs remain within their follicles in the ovary until they are fertilized. In most viviparous fish (those that bear hatched young) the eggs are shed into the oviduct before being fertilized. Moreover, although sperm are introduced into a female at copulation, the eggs are not fertilized until the following spring, the sperm remaining dormant until then. The eggs are small and contain little yolk, but the embryos into which they grow are nourished by a fluid given out from the ovary.

Once the fertilized eggs drop from their follicles into the cavity of the ovary, they develop rapidly, and the embryos grow a gill opening. Cilia (hairlike, beating organs) on this opening drive a current of liquid food through the gill and into the digestive tract, which at this stage is no more than a simple tube showing no sign of the future stomach and intestine.

Later, when the fins have developed, outgrowths from the dorsal and anal fins, which are rich in blood vessels, absorb nourishment. The liquid also contains the oxygen necessary for respiration. At a later stage some of the surface cells on the walls of the ovary drop away and are consumed by the growing embryos, as are any leftover sperm.

## Packed like sardines

The number of young varies with the species and with the size of the mother, but usually there are between 3 and 80. In the shiner or yellow-banded seaperch (*Cymatogaster aggregata*), for example, the number is 3 to 20, exceptionally 36. A mother 8 inches (20 cm) long would probably have 20 young, each about 1¼ inches (3 cm) long, tightly packed within the ovary.

## Fabulous fish

The first reports about the viviparity of surfperch were sent from California in 1853 to the celebrated Louis Agassiz, a professor at Harvard University. He published an account of these, calling them extraordinary fish. As so often follows from a fresh discovery, attention was focused on them, and they were written about in scientific journals the world over.

Even in the light of modern scientific discoveries, these facts about the surfperch remain fascinating. There are plenty of species of freshwater fish that bear fully hatched young, but very few marine fish. Few bony fish copulate, other than surfperch. Furthermore, the delayed fertilization of the ovum practiced by surfperch is uncommon, and it is unusual for embryonic development to take place in the ovarian cavity.

It is highly uncommon among viviparous fish for the embryos to be nourished other than

| SURFPERCH | |
|---|---|
| CLASS | **Osteichthyes** |
| ORDER | **Perciformes** |
| FAMILY | **Embiotocidae** |
| GENUS | **13 genera** |
| SPECIES | **24 species, including shiner perch, *Cymatogaster aggregata* (detailed below)** |

LENGTH
**Up to about 8 in. (20 cm)**

DISTINCTIVE FEATURES
**Similar to freshwater perch but with continuous dorsal fin; mainly silvery color, darker on back than on belly**

DIET
**Adult: small crustaceans, mollusks and algae. Young: mainly copepods.**

BREEDING
**Age at first breeding: 2 days; breeding season: summer; hatching period: 5–6 months; number of young: 3 to 36, born fully formed; male sexually mature from birth, female a couple of days after**

LIFE SPAN
**Not known**

HABITAT
**Shallows around eelgrass beds, piers and pilings in bays and quiet backwaters**

DISTRIBUTION
**Pacific coast of North America**

STATUS
**Common**

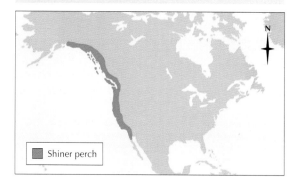

Shiner perch

through a yolk sac. There is also the unique situation of the young, or at least the young males, becoming sexually mature before birth. Perhaps it is this early maturity of the young surfperch that accounts for their being born in the surf—in a place likely to be reasonably free of natural predators at the time they are pairing.

# SURGEONFISH

MOST SURGEONFISH ARE very colorful, but they cause difficulties for those trying to name them because of their ability to change color. The name surgeonfish is easily explained: they carry lancets that can cut one's flesh as cleanly as a surgeon's scalpel.

Surgeonfish are deep-bodied and almost oval in outline except for the tail fin. They have small, rough scales and small gill openings. The tapering snout ends in a small mouth, which has a single row of teeth in each jaw, used for scraping food off coral. Both dorsal fin and anal fin start just behind the head and end just short of the tail. The pectoral fins are relatively large.

## Surgically enhanced

The lancets are small, extremely sharp, bony keels, one on each side of the body near the base of the tail fin. In some species they articulate like jackknifes, being hinged at the hind end. When not in use, these lancets lie in a groove or sheath. The lancets are used as weapons and can be rapidly erected and thrust out. Some members of the family have several lancets on each side, and anyone who handles the fish carelessly is liable to have their hands lacerated. Among the 58 species of surgeonfish, some grow to 2 feet (60 cm) in length. Their colors, though varied, have the subdued quality of pastel shades.

## Turncoats all

Surgeonfish live in shoals around coral reefs, so they are restricted to warm seas. They range from Madagascar east to Hawaii. They crop coral for small algae, eating a few small invertebrates living on the coral, though these are taken only accidentally. A common species is the yellow tang, *Zebrasoma flavescens*; it occurs in two color phases, one yellow and the other brown. When yellow, it is found mostly around Hawaii; when brown, it is seen throughout the Indian and South Pacific Oceans. The five-banded surgeonfish, *Acanthurus triostegus*, also found in the Indian and South Pacific Oceans, reaches

*A surgeonfish of the genus* Acanthurus *on the Great Barrier Reef, Australia. A deep, oval body outline is typical of the family Acanthuridae.*

*Shimmering blue fin markings frame the body of a transitional blue tang. The young of this species are yellow all over.*

## SURGEONFISH

| | |
|---|---|
| CLASS | **Osteichthyes** |
| ORDER | **Perciformes** |
| FAMILY | **Acanthuridae** |
| GENUS | **Surgeonfish: 6 genera; unicorn fish: 1 genus, *Naso*** |
| SPECIES | **Surgeonfish: 58 species, including blue tan, *Acanthurus coeruleus* (detailed below); unicorn fish: 16 species** |

LENGTH
**Up to 14½ in. (37 cm)**

DISTINCTIVE FEATURES
**Rounded body; spine near base of tail fin, which can inflict painful wounds. Adult: numerous deep blue longitudinal lines; spine is surrounded by yellow; concave tail without pale margin; dark margin on anal fin. Young: yellow overall, except for bright blue crescents above and below; transitional sizes may be part yellow, part blue (for example, blue fish with yellow tail).**

DIET
**Algae**

BREEDING
**Eggs float on surface, and larvae drift with plankton; when larvae approach suitable inshore habitat, they sink and develop into miniature of adult form**

LIFE SPAN
**A few years**

HABITAT
**Coral reefs, inshore weedy and rocky areas**

DISTRIBUTION
**Western Atlantic: New York and Bermuda south to Gulf of Mexico and Brazil; eastern Atlantic: waters around Ascension Island**

STATUS
**Common**

10 inches (25 cm) long, its dark apple-green body sporting dark brown vertical bars. It is also known as the convictfish. The blue tang surgeonfish, *A. coeruleus*, is blue when adult but bright yellow when immature, with a blue margin on the dorsal and anal fins. Marked changes in shape occur in some species, such as the humpnose unicorn fish, *Naso tuberosus*, which has a smooth forehead when young but develops a large bump on its head.

## Well-armed juveniles

The convictfish has a separate subspecies in the seas around Hawaii, recognized by a dark, sickle-shaped marking on the base of each pectoral fin. Most surgeonfish breed throughout the year, but the Hawaiian subspecies breeds only from December to July. Each female lays 40,000 eggs, which are about ¼ inch (6 mm) in diameter and contain an oil droplet that buoys them up to the surface. Each larva hatches in 26 hours and is just under ½ inch (2 mm) long. It floats upside down at the surface for another 16 hours, until the contents of its yolk sac are half used up. It then begins to sink gradually, and as it does, it starts to swim. At the end of four days, the larva has grown a swim bladder and is able to swim about and capture its food of plankton.

The elongated, almost tadpolelike larva changes after 2½ months into a diamond-shaped fish, flattened from side to side with long dorsal and anal fins. The second spine from the front of

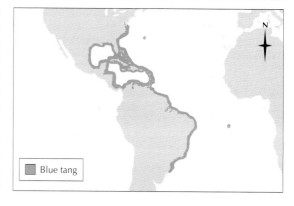

Blue tang

each fin is long, toothed throughout its length and venomous. The spine of the pelvic fin is similar in shape and also venomous. The early development of the convictfish is probably not typical, however, since the palette surgeonfish, *Paracanthurus hepatus*, of Florida to Brazil passes through the larval form in a matter of days.

A notable feature of the surgeonfish larva is the great change that takes place in the intestine. Whereas the larva feeds on animal plankton, the adult is wholly or almost entirely vegetarian. As the larva develops into a young fish and takes on something of the adult form, the intestine grows, extending by approximately three times its original length. This enables it to cope with the wholesale change in diet.

## Sting in the tail

While a surgeon's scalpel is wielded with good intent, the lancets on the tail of a surgeonfish are exclusively weapons of offense, comparable to the barb in the tail of a stingray. An outstanding feature of the stingray, the flat fish related to the skate, is that it swims with its pectoral fins, and the tail is reserved almost entirely for lashing to bring the sting into action. The surgeonfish, so different in shape, also swims with its pectoral fins and not with the tail, unlike other fish of its shape. Because it is mainly vegetarian, it does not require speed to catch food, and its formidable weapons of defense preclude any need for speed to get out of trouble. It swims with a rowing action of its pectoral fins, the tail fin at most making leisurely waving movements.

## Tolerant of cleaners

When threatened, a surgeonfish thrashes its tail violently from side to side, the lancets sticking out on each side and giving an adversary sharp, slashing cuts. As a result, other fish give it a wide berth, with the exception of certain kinds of cleaner fish, which are permitted to remove parasites from its skin. When a surgeonfish needs cleaning, it swims over to where the cleaner fish has its station and changes color. In the Atlantic ocean surgeon (*Acanthurus bahianus*), which has been studied closely, the surgeonfish changes from the normal reddish purple to a dark olive brown. It is probably a sign of peace, an indication that it will not slash its benefactor, which is nourished on the parasites that it nibbles from its dangerous client.

*A shoal of yellow tang on the Great Barrier Reef. Their scalpel-like lancets, for which surgeonfish are named, can be seen at the base of the tail.*

# SURINAM TOAD

*The female Surinam toad carries her eggs in pouches on her back for up to 140 days. During this time they metamorphose into tadpoles and may even develop into toadlets.*

THE SURINAM TOAD IS native to South America and the island of Trinidad in the Caribbean. It reaches a length of about 4 inches (10 cm), the male being smaller than the female, and has a flattened body. Its head is small and triangular and on the upper lip near the eyes are flaps of skin or short tentacles. Its skin is slippery and covered with small wartlike projections called tubercles. Like all members of its family, the Surinam toad has neither tongue nor teeth, and it is sometimes referred to as a tongueless frog. Those amphibians that spend much of their time on dry land use their tongue to moisten food before swallowing it. Because the Surinam toad has an almost completely aquatic lifestyle, and is able to remain submerged for up to 1 hour without rising to the surface for air, it has no need of such an organ. Although it has very small eyes, it can see in all directions, an advantage for any animal in detecting predators. The toad's feet have long fingers and toes but only the hind feet are webbed; it uses its front feet mainly to capture food and to push it into its mouth. The toad is a blackish brown color above, which camouflages it in the black mud of the streams and pools where it lives. Its underside is paler brown spotted with white or sometimes whitish with a dark brown stripe along the belly.

Surinam toads are sexually dimorphic (the male and female vary in appearance), albeit subtly. The female has a cleft between two fleshy bulges in the region of her cloaca (a chamber into which the urinary, intestinal and generative canals feed); the same region in the male is flatter.

Surinam toads live quite well in captivity. However, they do not often breed under the artificial conditions of an aquarium.

## Carrying her eggs on her back

The Surinam toad is noted chiefly for its remarkable breeding habits. Mating takes place soon after the start of the rainy season. After uttering a series of metallic ticking calls, the male grasps a female; if she is not ready to mate, the female quivers to indicate her rejection of the male.

During amplexus (the mating embrace), the conjoined female and male perform a series of somersaults in the water, which are initiated by the female. At the top of each somersault, at which point both male and female are on their backs, the female lays 3 to 10 eggs, which fall onto the belly of the male. The two continue through their descending arc, during which the male loosens his grip, allowing the eggs to roll onto the soft, spongy skin on the female's back, where he simultaneously fertilizes them. The eggs adhere only to the female's back, not to the male's belly or to each other. This procedure may be repeated up to 18 times and in all, 60 to 100 eggs may be laid.

When the last egg has been laid, the male swims away, but the female remains stationery. The skin of her back slowly swells and grows around each egg. A horny lid forms over the top so that each egg lies in its own pocket, or brooding pouch. At this point the whole of the female's back takes on a honeycomb-like appearance. Although the female now appears bigger, her enlarged state does not seem to inconvenience her in swimming and catching food.

As with other frogs and toads, the eggs hatch into tadpoles, but this takes place inside each small pocket, where the tadpole remains until metamorphosis is completed. During this period the tadpole is nourished by means of the many capillaries in its tail, which perform the same function as a placenta, plugging in to the mother's system in order to exchange nutrients and gases. The favorable conditions enable the tadpoles to metamorphose rapidly.

# SURINAM TOAD

| | |
|---|---|
| CLASS | **Amphibia** |
| ORDER | **Anura** |
| FAMILY | **Pipidae** |
| GENUS AND SPECIES | ***Pipa pipa*** |

ALTERNATIVE NAME
**Tongueless frog (name applied to all species in family Pipidae)**

LENGTH
**⁴⁄₁₀–⁷⁄₁₀ in. (10–18 mm)**

DISTINCTIVE FEATURES
**Extremely flattened body; large, triangular head; very small eyes; flaps of skin or short tentacles near eyes; fingers of hands terminate in star-shaped appendages; no tongue or teeth; flipperlike, webbed hind feet; blackish brown upperparts; paler brown underparts, spotted with white**

DIET
**Worms, insects, crustaceans and other aquatic invertebrates; also small fish**

BREEDING
**Number of eggs: up to 100. Skin on female's back thickens around each egg, creating individual brooding pouches. Hatching period: tadpoles or toadlets emerge after 90–140 days.**

LIFE SPAN
**Not known**

HABITAT
**Muddy bottoms of ponds and slow-flowing rivers, streams and canals**

DISTRIBUTION
**Ecuador east to Guianas, south to Peru, Bolivia and Brazil; also on Trinidad**

STATUS
**Common**

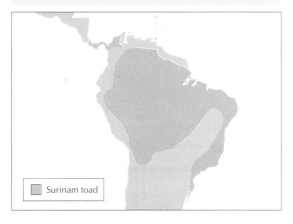
Surinam toad

About 3–5 months after conception, the lids of the pouches open and the tadpoles, or toadlets, depending on the degree to which they have developed, emerge and begin to swim around freely. The emergence of the young usually occurs when the mother sheds her skin. They generally escape by means of their own efforts, although the mother can exert pressure to force them to emerge if necessary. The young initially find diving problematic and remain near the surface of the water, but after about 1 month they are able to dive and swim.

## Skillful scavenger

The Surinam toad's flat, loose-skinned body looks most ungainly if it is taken out of water, but in water the toad swims strongly and gracefully, using its powerful hind legs. It lives its life almost entirely in the water, scavenging in the mud for any small aquatic invertebrates, dead or alive, that its long, slender fingers can sweep up. At the tip of each finger is a cluster of glandular filaments that are used to sense prey in mud or when in silty streams or rivers, where light conditions are poor.

The Surinam toad does not hibernate, but during dry weather it buries itself in the mud. Often, large numbers of other toad species bury themselves in the same area.

*Its flattened body shape and mottled brown color provides the Surinam toad with excellent camouflage, specially when foraging on murky river bottoms.*

# SWALLOWTAIL

*A tiger swallowtail, Papilio glaucus, of Canada and the United States. Swallowtails are generally large, colorful butterflies, named after the long tails on the hind wings of many species.*

SWALLOWTAIL BUTTERFLIES derive their name from the tails on their hind wings, which recall the forked tail of the swallow. They all belong to one family, the Papilionidae, but not all members of the family have tails. There is, however, another feature that is found in all members of the family. It is the possession by the larva of a protrusible Y-shaped organ, known as the osmeterium, situated just behind the head. It is usually colored orange or yellow and is normally hidden, but it can be suddenly pushed out if the larva is disturbed. The osmeterium is connected to glands in the thorax of the cater-pillar, and when pushed out it disseminates a strong scent that varies according to species. This organ is also present in the larvae of the Apollo butterflies (discussed elsewhere), which, although they are members of the family Papili-onidae, are not usually regarded as swallowtails.

The 600 species of swallowtails are extremely diverse. They are found everywhere except in the Arctic and Antarctic regions. Most species are tropical: North America has 20 to 30 species, whereas Europe has only five or six. One British species, *Papilio machaon*, has a wide distribution, ranging from western Europe across the whole of temperate Asia to Japan, and is represented by numerous subspecies.

Another swallowtail species, *Iphiclides podalirius*, is referred to as the scarce swallowtail on the basis of a few specimens captured in southern England. It is common in France, Germany and other parts of Europe. *P. hospiton* is confined entirely to Corsica and Sardinia.

## Varied habitats

Swallowtails range widely in their choice of habitat; races of a single species may have markedly different habitats. For example, the British race of the common swallowtail is confined to marshes and fens, and the French subspecies is an inhabitant of open country, par-ticularly chalk downs. In tropical rain forests many swallowtails fly high in the treetops. Some

## SWALLOWTAILS

| | |
|---|---|
| PHYLUM | **Arthropoda** |
| CLASS | **Insecta** |
| ORDER | **Lepidoptera** |
| FAMILY | **Papilionidae** |
| GENUS | **Many, including *Papilio*, *Parides*, *Eurytides* and *Iphiclides*** |
| SPECIES | **600, including common swallowtail, *Papilio machaon*; giant swallowtail, *P. cresphontes*; giant African swallowtail, *Drurya antimachus*; citrus swallowtail, *P. aegeus*; tiger swallowtail, *P. glaucus*; and ruby-spotted swallowtail, *P. anchisiades*** |

ALTERNATIVE NAME
**P. *aegeus*: orchard swallowtail**

LENGTH
**Wingspan: *P. machaon*, 2¾–4 in. (7–10 cm); *P. anchisiades*, 2⅖ in. (6 cm); *D. antimachus*, 10 in. (25 cm)**

DISTINCTIVE FEATURES
**Generally large with strikingly colored hind wings; tails on hind wings (most species); 3 pairs of well-developed legs (adult only)**

DIET
**Plant material; food plants vary according to species. *P. machaon*: milk parsley (genus *Peucedanum*). *P. aegeus* and others: cultivated plants of citrus and pea families.**

BREEDING
**Breeding interval: number of generations per year varies with both species and environment. *P. glaucus*: 1 to 3 generations per year.**

LIFE SPAN
**Many months in some species**

HABITAT
**Varies according to species, but most in tropical and temperate forests and grasslands; occasionally cultivated crops, marshes and hillsides**

DISTRIBUTION
**Most regions except Arctic and Antarctic. *P. glaucus*: exclusively North America. *P. machaon*: Europe east through temperate Asia to Japan; North Africa.**

STATUS
**Many species locally common**

*The swallowtails' life cycles vary according to species, habitat, food quality, temperature and distribution. Papilio machaon, of temperate Eurasia, is pictured.*

swallowtail species inhabit temperate forests and grasslands. Others can be found occasionally on cultivated crops or at high altitudes.

One feature of butterfly life in the rain forest, in which swallowtails feature prominently, is the habit of gathering in closely packed crowds on sand or gravel beside rivers and streams. If these butterflies are carefully observed, they may be seen to suck up moisture with their long tongues. Scientists have now established that most of these congregations gather where the urine of animals that come to drink at the waterside has soaked into the sand. The butterflies seem to be attracted by the organic salts present in the urine. They also gather at mineral springs and wet ashes. Such gatherings consist only of male swallowtails; the females feed on the nectar of flowers and seldom come down to ground level.

## Snakelike and poisonous caterpillars

Swallowtails have a typical butterfly life history. The pupae are attached by the tail to a leaf or twig. They face head upward and are supported by a silk girdle. The larva feeds on the leaves of plants; food plants of the common swallowtail are milk parsley and fennel in the wild, but it can be reared in captivity on carrot leaves. One group of mainly tropical species confines itself to the poisonous creepers of the pipe vine, genus *Aristolochia*. The poison taken in from this plant renders the caterpillars inedible to birds, and as it persists in the pupa and the adult insect, these also are protected against predators.

Another form of protection in many swallowtail larvae is provided by the eyespots on the forepart of the body, which give the larva a snakelike appearance. In the Tropics, tree snakes are among the most deadly enemies of lizards and small birds. Merely to be reminded of a snake may well be enough to alarm the birds and lizards and so effectively discourage attack. The scent of the osmeterium is undoubtedly defensive in function and may be directed against parasitic wasps and flies.

## Mimicry as defense

Many swallowtails profit by having the appearance of other, less palatable species. For example, the poisonous *Aristolochia*-feeding species are protected from predation because they have a distinctive appearance and birds quickly learn to recognize them as poisonous. In the course of evolution, other swallowtails, which are usually acceptable as food for birds, have come to resemble the poisonous species and thereby gain protection by using mimicry.

One of the best-known mimetic species is *P. memmon* of East Asia. The male is a large bluish black butterfly that shows little tendency to vary. However, the female appears in a number of varieties or forms. Some possess tails; others do not. Each of the forms is a copy of one of the species of *Aristolochia* swallowtails. Another well-known mimic is the African swallowtail, *P. dardanus*. In this species too, only the females are mimetic (the males are colored plain black and pale yellow) and, like the females of *P. memmon*, they adopt a multiplicity of forms. However, none of these forms mimics other swallowtails. Instead, they have the resemblance of butterflies from various other families that predators regard as distasteful.

In most parts of Africa, a female *P. dardanus* is usually a mimic of one of the poisonous species occurring in the locality. Some female forms of *P. dardanus* mimic monarch butterflies of the genus *Danaus*. Only in Ethiopia and Madagascar is this butterfly nonmimetic, and in these regions the females resemble the males.

## Glorious butterflies

The swallowtails include some of the most brilliantly colored and striking of all butterflies. Lepidopterists consider that the finest of them are the birdwings of Southeast Asia, New Guinea and tropical Australia. These butterflies are described elsewhere. The female of one, *Ornithoptera alexandrae* from New Guinea, has a wingspan of up to 11 inches (28 cm) and is the largest known butterfly. It is also a rare species and is now protected. The giant African swallowtail, *Drurya antimachus,* native to tropical western Africa, is another enormous swallowtail butterfly. It is orangish brown with black markings and has a wingspan of up to 10 inches (25 cm). The male is much larger than the female, a condition contrary to that seen in large birdwings.

The species that swarm on sand in tropical Asia are mostly kite swallowtails of the genus *Graphium*, many of which are black barred or spotted with brilliant green or blue colors. The Bhutan glory, *Armandia lidderdalei,* of Assam, and the royal swallowtail, *Teinopalpus imperialis*, which is found in the same region, both have three tails on each of their hind wings. The small Asian swallowtails, *Leptocircus,* have tails that are longer than the span of their forewings and their flight is as rapid and as strong as the flight of a hawkmoth.

*Swallowtails have large hind wings, making them very strong fliers. Pictured is the five-bar swallowtail,* Graphium antiphates itamputi, *from Indonesia.*

# SWALLOW-TANAGER

THE SWALLOW-TANAGER has sometimes been classified along with
the true tanagers. At other times,
ornithologists have considered it sufficiently different in form and habits
from true tanagers to be classed separately. About 5½–6 inches (14–15 cm)
long, it has longer wings and shorter
legs than other tanagers. It has a swallowlike bill, which is broad with sharp
cutting edges and a hook at the tip of
the upper mandible. The swallow-
tanager has been described as one of
the most colorful birds of the American
Tropics. The male is mostly turquoise
blue when seen against the light, but
this changes to emerald green when
seen with the light. The face and throat
are black with black bars on the flanks
and white in the center of the belly. The
female is very different in appearance;
she is bright green with yellowish
underparts and a gray throat.

Swallow-tanagers range across
Central America and South America, from
Panama to Bolivia and Brazil. They are also
found on Trinidad in the Caribbean.

## Partial migration

Outside the breeding season, swallow-tanagers
live in small flocks of up to 12 individuals,
keeping to themselves. At that time of the year
they are found in the humid lowlands and, less
frequently, in montane areas. However, at the
start of the breeding season they move into
mountainous and hilly regions, particularly
where there is evergreen forest festooned with
dense growths of creepers. The lower limit of the
swallow-tanagers' breeding range is at the level
of the deciduous woodland.

Swallow-tanagers are not as vocal as many
other tropical birds. Their songs are very weak
and are not heard until late in the morning. A
typical noise made by swallow-tanagers is a
high-pitched *tsee*. The bird repeats the sound
every 5 to 6 seconds.

## Soft food preferred

Swallow-tanagers feed on both fruits and insects.
Fruit forms the bulk of their food in the dry
season, but they turn to insects at the start of the
rainy season, when swarms of flying insects
appear. These are caught flycatcher fashion, by
flying out from a perch as an insect passes.
Swallow-tanagers catch mostly flies, but they
also eat small locusts, butterflies and flying ants.
They prefer soft-bodied insects. This is not
surprising, because swallow-tanagers also prefer
soft, pulpy fruit such as avocados.

## Mass curtseying

The display of the swallow-tanager is simple but
repetitive. Male and female display to each other,
the female taking the initiative by curtseying,
bobbing up and down with the body rigid. The
pair curtsey out of phase so that the male is
upright when the female is crouched. The bobbing may be repeated up to 100 times. Birds of
the same sex sometimes display toward each
other using similar curtseying movements.

Swallow-tanagers differ from the true
tanagers by nesting in holes rather than building
cup-shaped nests in the open. They may use
crevices in buildings and the abandoned
burrows of jacamars or puffbirds. Alternatively,
the swallow-tanagers may excavate their own
burrows in a bank, which vary from as little as
2 inches (5 cm) to as much as 6½ feet (2 m) deep,
and 3⅓–10 feet (1–3 m) above the ground. When
excavated in earth, the burrows have entrance
holes 2⅖–4 inches (6–10 cm) in diameter. The nest
of grass and flowers is built at the end of the
burrow by the female, who also digs the hole.
Although female swallow-tanagers do most of
the work, the males may help with the collection
of nesting material. The three white eggs hatch in

*The swallow-tanager's
bill is ideally shaped
for its preferred diet
of soft fruits and
insects. It is broad,
hooked at the tip and
has sharp edges. A
captive male is shown.*

*Swallow-tanagers (male, above) choose high regions when breeding, and avoid the lowlands.*

13–17 days. The male helps to feed the chicks, which stay in the nest for 24 days. The males sometimes try to induce the females to attempt a second brood, with infrequent success. Male swallow-tanagers acquire adult plumage over four years, but breed in subadult plumage.

## Gluttons for fruit

The name swallow-tanager could almost be a slang term describing the feeding habits of these birds. Biologist Ernst Schaefer, who has studied swallow-tanagers in the wild and in captivity, has described the way the large, sharp-edged bill is used to scrape the fleshy pulp from fruits and then eject the stone instead of swallowing it, as many other fruit-eating birds do. The mouth is opened very wide and the fruit, perhaps larger than the bird's head, is engulfed in the manner of a snake working its way over its prey. The fruit is then twisted around in the mouth until all the flesh is scraped off.

Schaefer's captive swallow-tanagers ate two-thirds of their own weight in fruit each day. The birds do not have a crop but store the pulp in a greatly elastic throat, which bulges out.

---

# SWALLOW-TANAGER

| | |
|---|---|
| CLASS | **Aves** |
| ORDER | **Passeriformes** |
| FAMILY | **Thraupidae** |
| GENUS AND SPECIES | ***Tersina viridis*** |

WEIGHT
**About 1 oz. (29 g)**

LENGTH
**Head to tail: 5½–6 in. (14–15 cm)**

DISTINCTIVE FEATURES
**Relatively long wings; broad, flat bill. Adult male: mostly bright blue plumage; black forehead and throat; white belly. Adult female: mostly green plumage; gray throat; yellow belly; green-barred yellow flanks. First-year male: mottled green and blue; no black on face.**

DIET
**Soft fruits, berries and insects, taken on the wing**

BREEDING
**Age at first breeding: probably 2 years; breeding season: February–June (northern South America), September–November (southern Brazil); number of eggs: 3; incubation period: 13–17 days; fledging period: 24 days; breeding interval: 1 year**

LIFE SPAN
**Not known**

HABITAT
**Open woodland, forest edge, secondary growth and strips of riverside forest**

DISTRIBUTION
**Panama south through northern South America, east of the Andes to southern Brazil; also on island of Trinidad**

STATUS
**Locally common**

Swallow-tanager

# SWAMP EEL

*Though unrelated to its namesake, the swamp eel exhibits some of the habits of eels, notably in its ability to wriggle over dry land to reach new watery habitats.*

SWAMP EELS are unrelated to true eels. They mainly breathe air, having tiny gills. The fins are little more than ridges. The eyes are covered with skin or are absent. There are about 16 species, found in the swamps of the American Tropics, West Africa, Asia and Australia. The blind cave eel, *Ophisternon candidum*, inhabits underground waters in Australia. The cuchia, *Macrotrema caligans*, lives off Malaysia.

Swamp eels live in oxygen-poor waters. They shun the light, moving mainly at night. Rather than breathe through gills, they rise to the surface to gulp air or wriggle out over mud. Some species are amphibious, like frogs and toads. In summer swamp eels burrow into the mud and go into a summer sleep, or estivation. The cuchia digs down as deep as 2–3 feet (60–90 cm). Its eyes are specially adapted to cope in the mud. They are covered with a thick, semitransparent skin that is flush with the body. The eyes are of little use. Even those swamp eels with relatively efficient eyes can do no more than distinguish light and dark, and several species are blind.

## Hearty appetites

If a swamp eel's eyes, gills and fins are near to useless, the mouth makes up for them. It is large and thick-lipped, with rows of teeth on the jaws and palate. A swamp eel eats small invertebrates, such as worms and snails, and fish. A rapacious animal, it can eat its own weight of food in a day.

## Bubble-nesting

One species, *Monopterus albus*, has been found to be first male, later changing to female. Before spawning, the male builds a nest of bubbles. He gulps air at the surface and spits out the mucus-covered bubbles at the surface to form a raft. As the female lays, the male takes each egg in his mouth and spits it onto the raft. He then cares for the eggs, supplying them with oxygen through his skin. When the fry hatch, they have large pectoral fins well supplied with blood vessels that are used for breathing. These drop off when the young fish are 10 days old. Males become female within 8–30 weeks.

# SWAN

THE SEVEN SPECIES OF swans are closely related to geese. Together they make up the tribe Anserini, which belongs to the subfamily Anserinae, part of the order Anseriformes. Swans look much alike in terms of structure. However, the South American coscoroba swan, *Coscoroba coscoroba*, which is the smallest swan and has a comparatively short neck, is somewhat different from the other six species (genus *Cygnus*). The coscoroba swan has several characteristics in common with the whistling or tree ducks (tribe *Dendrocygnini*).

## Familiar, much-loved birds

The most familiar swan is the mute swan, *Cygnus olor*, which originally was native to parts of Europe and Asia but has been domesticated and introduced into many parts of the world, such as North America and Australia, where it has gone wild. It is thought that the species was introduced into Britain by the Romans. The mute swan is about 5 feet (1.5 m) long. Its plumage is all white and the bill is bright orange with a prominent black knob at the base. Bewick's swan, *Cygnus columbianus bewickii*, and the whooper swan, *C. cygnus*, are the other two swans that breed in Eurasia. Bewick's swan breeds in the marshy tundra of northern Russia and Siberia and migrates south to temperate wetlands in Europe and Central and eastern Asia for the winter. The whooper swan breeds farther south, including in northern Scandinavia and Iceland, with a few pairs nesting sporadically in Scotland. Both types of swans have black bills with a yellow base, the pattern differing slightly between the two. Bewick's swan is rather smaller than the whooper, with a shorter neck.

Two swans are found in North America. The tundra swan, *Cygnus columbianus columbianus*, is the American counterpart of Bewick's swan, and was formerly known as the whistling swan (tundra and Bewick's swans are subspecies, or races, of the same species). The tundra swan has much less yellow on the bill than its Eurasian

*Swans (mute swan, below) defend their cygnets aggressively, attacking anything that comes too close for comfort.*

# TUNDRA SWAN

| | |
|---|---|
| CLASS | **Aves** |
| ORDER | **Anseriformes** |
| FAMILY | **Anatidae** |
| GENUS AND SPECIES | ***Cygnus columbianus*** |

ALTERNATIVE NAME
**Bewick's swan (Eurasian subspecies, *C. c. bewickii*); whistling swan (former name for American subspecies, *C. c. columbianus*)**

WEIGHT
**9½–21⅔ lb. (4.3–9.6 kg)**

LENGTH
**Head to tail: about 4⅓ ft. (1.3 m); wingspan: 6½–7 ft. (2–2.2 m)**

DISTINCTIVE FEATURES
**Very long neck, held vertically with a kink at bottom; black legs and feet. Adult: pure white plumage; small yellow spot of variable size at base of bill (*C. c. columbianus*); entire upper half of bill yellow (*C. c. bewickii*); rest of bill black. Juvenile: gray overall; pink bill.**

DIET
**Mainly aquatic plants; also grasses and crops**

BREEDING
**Age at first breeding: usually 3–4 years; breeding season: eggs laid late May–June; number of eggs: 4 or 5; incubation period: 30–32 days; fledging period: 60–75 days; breeding interval: 1 year**

LIFE SPAN
**Up to 25 years**

HABITAT
**Summer: small lakes and ponds in tundra; winter: estuaries, marshes and shallow lakes**

DISTRIBUTION
**Breeds across tundra of far north; migrates to localized winter sites in temperate latitudes**

STATUS
**Locally common**

Tundra swan ▨ summer ▢ winter

relative: no more than a yellow spot at the base. It breeds mainly north of the Arctic circle and migrates to southern Canada and the United States for the winter. The second North American swan, the trumpeter swan, *C. buccinator*, used to breed over much of Canada and the United States but it is now confined to Alaska, southwestern Canada, the northwestern United States and parts of the Midwest. In 2000 the total wild population was only about 13,500. Trumpeter swans resemble the slightly smaller tundra swans but lack the yellow spot on the bill, and the eyes are almost totally enclosed by black.

The only swans in the Southern Hemisphere, apart from the coscoroba swan, are the black swan, *C. atratus*, of Australia, and the black-necked swan, *C. melanocoryphus*, which is found from Brazil south to Tierra del Fuego and the Falkland Islands. The black swan is all black but with white primary wing feathers and a bright red bill. It has been introduced into New Zealand. The black-necked swan has a black head and neck, a white eye stripe and a red bill.

*A pair of trumpeter swans, Kenai Peninsula, Alaska. The trumpeter swan, which is one of the world's heaviest flying birds, suffered a dramatic population decline during the last century.*

*Black swans are highly nomadic, flying large distances across the arid inland plains of Australia in search of temporary wetlands. Flocks of hundreds or even thousands of black swans gather at rich feeding grounds.*

## Not so mute

Compared with other swans, the mute swan is quiet, but its name is a misnomer because it has a variety of calls. A flock of mute swans can be heard quietly grunting to each other as they swim along a river or across a lake. When disturbed or in defense of the nest, mute swans hiss violently. The sighing noise during flight is caused by the wings.

The whooper swan has a buglelike call when flying and a variety of quiet calls when grounded. Bewick's swan has a pleasant variety of honks and other sounds, and the trumpeter swan is named after the trombonelike honking calls produced in the long, coiled windpipe. It is said that the swan song, the legendary song of a dying swan, is based on a final slow expiration producing a wailing noise as it passes through the long windpipe.

## Heavyweights of the air

Despite their great weight, swans are strong fliers. They have four times the wing loading (the body weight divided by the surface area of the wings) of a herring gull, *Larus argentatus*, or American crow, *Corvus brachyrhynchos*, and they have to beat their wings rapidly to remain airborne. A high wing loading makes takeoff and landing difficult, and swans require a long stretch of open water over which they can run to gain flying speed or surge to a halt when landing. Swans are also unable to maneuver in flight, and one of the chief causes of mortality in built-up parts of the world is collision with overhead power cables.

## Shallow water feeders

Swans feed mainly on plants but they also feed on animals such as tiny fish, tadpoles, insects and mollusks. They often feed on land, grazing on grasses, herbs and sedges like geese, but more often they feed on submerged aquatic plants, which they may collect from the bottom by lowering their long necks underwater, sometimes upending like ducks. This limits the swans' distribution to stretches of shallow water because they very rarely dive and are only occasionally seen on deep water.

## Aggressive in defense of the nest

Swans nest near water. Male mute swans set up territories, each defending a stretch of river from which they drive other males and young swans. Intruders are threatened by an aggressive display in which the neck is drawn back, the wings are arched over the back and the swan propels itself in jerks with the webbed feet thrusting powerfully in unison, instead of alternately as in normal walking. There are a variety of displays between the cob (male) and the pen (female), involving tossing and swinging the head and dipping it into the water.

Mute swans mate for life and nest in the same territory each year, some violent fights taking place if a new pair tries to usurp the territory. The nest is a mass of water plants and twigs, roughly circular and cone-shaped with a depression in the center. Wild mute swans nest among reeds on small islands in pools, but semi-domesticated ones may nest on the banks of ponds in parks or in other inhabited places. Mute swans occasionally nest in colonies rather than in spaced-out territories, such as the large colony at Abbotsbury in southern England. There usually are about five eggs, rarely up to twice as many,

and they are incubated mainly by the female, the male taking over only when she leaves to feed. In the smaller swans incubation lasts about 4 weeks, but it is 5 weeks in the larger species and 5½ weeks in the black swan. While the last eggs are being brooded by the female, the male takes the cygnets (young) to the water. The family stays together until the cygnets fledge at 4–5 months. When young, they swim together in a tight bunch with the female leading and rooting up plants for them to eat.

## Swan-upping

By the 13th century the mute swan no longer existed as a wild bird in England. All swans were the property of the Crown or certain individuals and bodies who owned swans under royal license. The sovereign had a swan master, who enforced the practices of swan keeping. Individual swans were marked on the bill or feet with a series of notches or more elaborate marks to indicate ownership. This practice still survives on the Thames River, where cygnets are marked annually in a practice known as swan-upping, and the Abbotsbury swans are still marked on the webs of the feet.

*In spring, Bewick's, tundra, whooper and some trumpeter swans fly north to taiga or tundra regions to breed. They nest on small ponds or lakes. Pictured are two pairs of whooper swans.*

# Index

Page numbers in *italics* refer to picture captions.
Index entries in **bold** refer to guidepost or biome and habitat articles.

Page numbers in *italics* refer to picture captions. Index entries in **bold** refer to guidepost or biome and habitat articles.